W0018095

A MINE TO MAKE A MINE

Environmental Management

A Mine to Make a Mine

Financing the Colorado Mining Industry, 1859–1902

By JOSEPH E. KING

TEXAS A&M UNIVERSITY PRESS

College Station and London

Copyright © 1977 by Joseph E. King
All rights reserved

Library of Congress Cataloging in Publication Data

King, Joseph E   1942–
    A mine to make a mine.

        Bibliography: p.
        Includes index.
        1. Gold mines and mining—Colorado—Finance—
History. 2. Silver mines and mining—Colorado—Finance
—History. 3. Mining industry and finance—Colorado—
History.
    I. Title
    HD9536.U53C64   338.2'3   76-51655
    ISBN 0-89096-034-8

Manufactured in the United States of America
FIRST EDITION

To Claire

Contents

List of Illustrations

Abbreviations

AJM   *American Journal of Mining, Milling, Ore Boring, Geology, Mineralogy, Metallurgy, etc.* (New York), 1866–July, 1869.

DAB   Allen Johnson and Dumas Malone, eds. *Dictionary of American Biography.* 20 vols. New York, 1928–1936.

EMJ   *Engineering and Mining Journal* (New York), 1869–1902.

MSP   *Mining and Scientific Press* (San Francisco), 1860–1902.

Preface

Between the Civil War and the turn of the century, the gold and silver mines of Colorado were a gaudy and unsavory, though important, element of the American financial scene and in the economic history of the Rocky Mountain West. During those forty-odd years, country folk, small-town merchants, big-city businessmen, politicians, and a wide cross section of the American public organized mining companies and bid for mining shares in a feverish, usually fanciful race to make a fortune in the mountains of Colorado. Word that gold had been discovered near Black Hawk in 1859 aroused the nation's curiosity and whetted its appetite for easy wealth. Over the next four decades, the influx of capital to work the mines transformed a remote and forbidding region into an industrial province on the frontier and an acknowledged leader among mineral producers in the United States. By 1902, Cripple Creek, the last of the great bonanza camps in the state, experienced a sharp decline in production and signaled the end of an era. The boom days were over; for the mines, the future held in store leaner grades of ore, more conservative business and financial practices, governmental regulation, and stronger competition for the investors' dollar.

By that time, however, entranced by the same vision of instant wealth that kept the fabled prospector and his trusty burro combing the hills for the one huge deposit of pay ore, American investors had poured millions of dollars into Colorado in hopes of striking it rich. Not surprisingly, the prospector and the lusty boom towns he visited have captivated the imagination of historians at the expense of the later stages in the development of a mineral industry. Yet, typically lacking the means to make his discovery a paying mine, the prospector transferred his claim to a promoter who undertook a new search for wealth—the capital funds needed to open the vein in depth and excavate the ore. Starting with the work of the promoter, the history of the western mines took on a new shape, characterized by clever advertising, schemes to raise funds on eastern money markets, and efforts at combining the men, machinery, and management to turn a hole in the ground into a profitable enterprise. Colorado, it was quickly learned, was not to be a poor man's paradise where an individual with pan and pickaxe could carry off its mineral resources. Quite the contrary, since the seams of gold and silver were encased by hardrock and buried deep within the mountains, the area would challenge America's entrepreneurial skill and the commitment to acquiring sudden riches.

Thousands tried. Bolstered by the mischievous myth that mining simply involved digging hidden treasure from the earth, the precious-metal mines became the objects of lively and extensive speculation in the United States and, in the process, seduced the venturesome in all sections and economic groups. A relatively small expenditure bought an investor a chance at a kingly fortune. The idea, ceaselessly peddled by the promoting fraternity, that anyone could become a millionaire in the mines, or at least try, struck a responsive chord in a democratic, optimistic society, and the gullible inhabitants of small country towns joined with usually hard-headed members of the eastern business establishment to take a flyer in the mines. As a result, a mixture of business enterprises could be found in the mineral regions; single proprietorships and partnerships were common at all times, but the large capital requirements of deep-level mining fostered the spread of the corporation. The sale of securities on the public market enabled companies to raise more capital but minimized the risk to individuals. One visitor to the Rockies in 1882 revealed the impact of organized capital by pronouncing Colorado a "state of corporations," even as portions of the region were still menaced by hostile Indians.[1]

Indeed, the formative years of the mining industry in Colorado were plagued by numerous difficulties. The precipitous terrain, the harsh climate, and the rootless character of life on the frontier, along with the great distance separating the investor from his enterprise,

[1] John F. Graff, "*Graybeard's*" *Colorado: or, Notes on the Centennial State*, p. 41.

caused major problems. These conditions combined to hinder sound development of the mines by jeopardizing communications, multiplying expenses, and accentuating the risks inherent in the business of mining. These matters aside, mine owners also faced the sudden disappearance of the pay streak, cave-ins, underground floods, fires, and, the bane of miners working in rich ground, legal and extralegal disputes over ownership of the mineral. Finally, in an age marked by few restraints on business practices, investors were victimized by fraudulent promoters, worthless "experts," and stock-market manipulators, who gave the industry its corrupt reputation.

This then is a study of the process and problems of financing a precious-metals mining industry in Colorado during the nineteenth century. It is neither a story of the mining rushes nor a history of foreign investment in the West (the latter admirably treated by Clark C. Spence, *British Investments and the American Mining Frontier, 1860–1901*, and W. Turrentine Jackson, *The Enterprising Scot: Investors in the American West after 1873*). Rather, it is concerned with explaining the mobilization, migration, and impact of American, principally eastern, capital on a major state in the mining West. Moreover, Colorado's heavy reliance on distant sources for investment capital and business know-how typified the region after the Civil War. If success were to crown the efforts of mining men in this remote and speculative industry, it would depend upon the relationship formed between East and West, frontier and developed area, capitalist and promoter, luck and skill, and even dream and reality. In the final analysis, the reader must decide whether the evolution of the mining industry was, as one student said, an "instructive romance," or, as he later lamented, a "tawdry fiction" in which "unlimited debauchery" was allowed to "masquerade as romance, and the piling of millions by unscrupulous gamesters is not infrequently glorified into masterful finance."[2]

2 Thomas A. Rickard, "The Development of Colorado's Mining Industry," *Transactions of the American Institute of Mining Engineers* 26 (1896): 834; Thomas A. Rickard, ed., *The Economics of Mining*, p. 330.

Acknowledgments

THIS book could not have been completed without the generous help of many people. For their indispensable cooperation, I acknowledge with gratitude the staffs of the University of Illinois Library, the Denver Public Library, the State Historical Society of Colorado, the Baker Library of Harvard University, The Filson Club, the New York Public Library, the Chicago Historical Society, the New Mexico State Records and Archives Center, the New York Historical Society, the Maryland Historical Society, and the Wisconsin State Historical Society. I am grateful to John A. Brennan of the Western Historical Collections, University of Colorado, for his good and genial advice and, especially, to Gene M. Gressley of the Western History Research Center, University of Wyoming, for always responding generously to my appeals for help. A special note of thanks also to the late Donald M. Hyman, whose friendship will not be forgotten.

I want to express my appreciation to the Lincoln Educational Foundation for its John E. Rovensky Fellowship, which provided financial assistance. My thanks to Joan Weldon of the History Department at Texas Tech University for typing the manuscript. To my mentor at the University of Illinois, Professor Clark C. Spence, who continually interrupted his crowded schedule to give me wise and understanding advice, I owe my greatest debt of gratitude. Finally, to my wife Claire, though she be last she should be first, for her cheerfulness and patience as my chief critic, typist, and proofreader, I say thank you. Without her encouragement and support, none of this would have been possible.

J.E.K.

A MINE TO MAKE A MINE

1

A Booming Failure

No crowds of excited brokers burst into the exchanges of New York, Philadelphia, or Boston on May 6, 1859, to bid for shares in the new industry two thousand miles to the west. Nor did businessmen gather around mahogany desks or meet over lunch in posh restaurants to argue the pros and cons of buying a mine in the Rocky Mountains. In fact, the great financial centers of the East were quite unaffected by what mining engineer Thomas A. Rickard later heralded as the "birthday" of the Colorado mining industry.[1]

Communications being slow, many weeks would pass before the eastern seaboard learned that John H. Gregory, a Georgia-born prospector, had located the outcrop of a gold vein in Colorado and that mining in the Rockies was about to enter a new phase. Prior to Gregory's discovery in a narrow gulch of North Clear Creek, where later rose the town of Black Hawk, the then mountainous fringe of Kansas Territory had attracted in 1858 and 1859 hundreds of gold seekers who concentrated on removing the golden specks found in the many creeks flowing out from the base of the mountains. This elemental form of recovering the precious metal, called placer mining, was of short duration. As the areas of known value along the streams became overcrowded, the more experienced prospectors, among them Gregory, trekked deeper into the mountains in search of the weathered and heavily decomposed veins of ore hidden on the sides of pine- and spruce-covered ravines. Throughout the spring and summer of 1859, prospecting parties along branches of South Clear Creek, near Idaho Springs and Georgetown, and along Boulder Creek to the

[1] Thomas A. Rickard, "The Development of Colorado's Mining Industry," *Transactions of the American Institute of Mining Engineers* 26 (1896): 836.

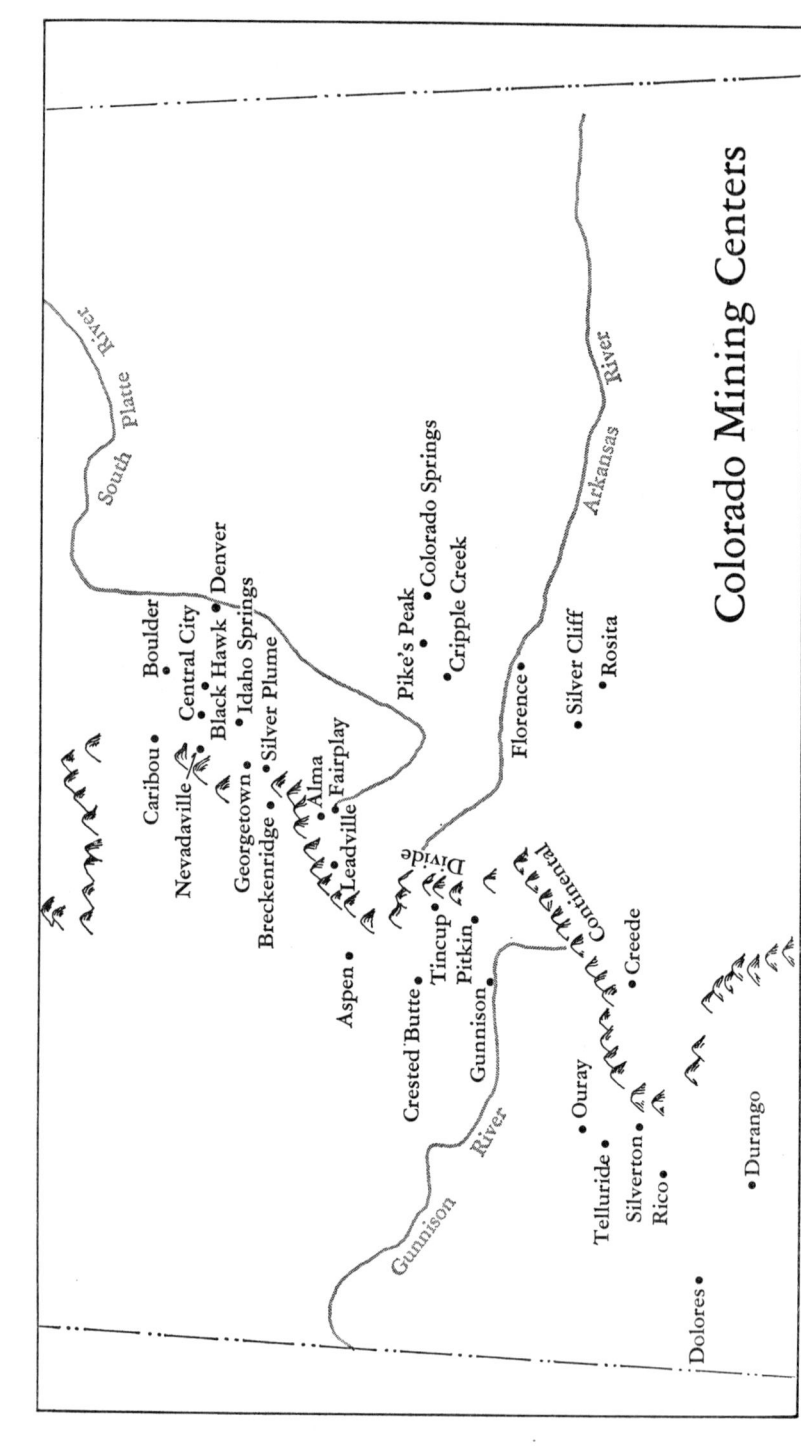

Colorado Mining Centers

north reported strikes of "blossom rock" similar to the one in Gregory Gulch.

Once Gregory and his associates had wisely laid claim to the lode by marking off its boundaries, they devoted most of their time and energy to making it pay. The soft surface dirt and decomposed vein matter were easily excavated by hand and the ore carried to the stream for washing in order to separate the metal from its matrix. The first pan yielded four dollars, and by the end of May 6 the group was richer by forty dollars in the form of a few ounces of gold. Subsequently, Gregory and those who rushed into the district behind him improved their techniques for treating larger quantities of ore. Crudely fashioned wheelbarrows were used to bring heavier loads to stream side, and rockers, sluices, and long toms, each an essentially trough-shaped wooden box with cleats or riffles at the bottom to catch the metal, hastened the task of washing. At some camps miners built flumes to increase the supply of water near the diggings and erected slides to move the ore downhill. Every one of the devices had been widely known in the West and was readily constructed from timber cut on surrounding hillsides.[2]

The passage of time inevitably brought more difficult mining conditions and greater complexity in working the lodes. As details of the excitement filtered eastward in 1860, new throngs of Argonauts were inspired to test their luck at Gregory Gulch and elsewhere; by way of illustration, the Recorder of Spanish Bar District at Idaho Springs alone registered 12,530 lode claims in the year beginning June 1, 1860. Typical of these stampedes, most of the locations proved worthless, except in the roseate hopes of the claimant, and irritated the "old timers" who scoffed at the multitude of shallow gopher holes sunk by newcomers on every available piece of ground. Aside from the swelling population, more vexing problems appeared as the miners exhausted the eroded surface deposits and encountered solid quartz, at depths of approximately thirty to forty feet, which had to be broken and crushed to free the gold. From the outset of work on the lode deposits there had been a few "knowing cusses" who foresaw the need for crushing machines and had lugged such equipment into the mountains from other mining regions, often

2 A good discussion of these early mining methods can be found in Percy S. Fritz, *Colorado: The Centennial State*, pp. 152–61.

California, or had begun to construct them once the need became obvious. An early favorite was the arrastra, borrowed from antiquity, which, powered by water wheel, mule, or ox, dragged heavy stones over the quartz dumped in its circular track. Prominent too were the primitive stamp mills operated by man or beast that, by pivoting a hammer, commonly a large boulder, on a tree trunk, ground the ore shoveled into a wooden box—a "woodpecker mill" one miner called his. If it seemed to work, no matter what it was called or where it came from, chances were great that neighbors would proclaim it invaluable and set to building one of their own, spawning further imitations throughout the camp.[3]

By far the most advanced piece of technology to enter Colorado in 1859–1860 was the steam-driven stamp mill that began to displace the arrastra in larger operations. Irregular or short supplies of water made steam power especially welcome in the mountains then as well as later in the mining development of the territory. Hazy though the specifics are on these early machines, it is clear that most were brought by eastern immigrants who had them manufactured at home and laboriously transported cross country at great expense. One of these made in Rockford, Illinois, and called the Black Hawk Mill had been shipped to the Missouri River, hauled by wagon more than six weeks over the plains, and was supposedly responsible for the name of the town that sprang up near it on Clear Creek. Leaving his infant sleeping-car business, George Pullman, with a party of Chicago men, brought a mill into Russell Gulch and erected it a few rods from their cabin, which then trembled to the "old engine puffing and stamps pounding on the quartz night and day only stopping to rest on Sundays."[4]

In fact many mills were stopping entirely. Despite Pullman's belief that the whistles and noise in the mountains meant a hundred steam engines at work, creating a commotion to equal the "infernal

[3] Ovando J. Hollister, *The Mines of Colorado*, p. 117; Frank Hall, *History of the State of Colorado*, 1: 204.

[4] There were several claimants to the honor of constructing the first mill in Colorado; for one see Henry D. Teetor, "The First Quartz-Mill in Colorado: Honorable Thomas J. Graham," *Magazine of Western History* 13 (November, 1890–April, 1891): 563. Donald C. Kemp, *Colorado's Little Kingdom*, pp. 48–49; Pullman to mother, September 28, 1860, George M. Pullman Papers, Chicago Historical Society.

region," the number of operating mills actually declined by the end of 1860 due to a new source of difficulties with the ore and its treatment that baffled even experienced miners. According to the men in the camps, the ore became "refractory" or "rebellious" at depths of fifty to one hundred feet, and the standard practice of crushing the ore in the mills, mixing it with mercury (for which gold has a natural affinity), and retorting the amalgam, or heating it to dispel the mercury and to leave the precious metal, was no longer effective. This tried and true method was failing because, simply stated, the character of the gold veins, which near the surface had been free except for the encasing rock, changed in depth to auriferous iron and copper pyrites, that is, sulphurets (the modern term being sulfides), which resisted treatment at the mills. Much to their chagrin, the miners discovered that they were losing gold in the reduction process. Some mills could save a mere 10 to 15 percent of the gold content, while most could do no better than remove 50 percent of the assayed value besides losing all of the base metals such as copper and lead. An operation on Gold Hill showed the suddenness of this development; the mill that had been yielding over $1500 in gold each week ran iron pyrites continuously for eighteen hours and recovered a paltry $1.40. In September, 1860, Pullman drily concluded that there was a large proportion of gold mined that was "entirely worthless" and estimated that one mill in ten paid expenses; he revised his figures to one in twenty the following month.[5]

Also taking their toll were poor judgment and inexperience. Novices erected mills before turning over a spade full of dirt, misused their equipment without the knowledge to repair it, and violated basic rules of reduction at nearly every stage of the process. There was, for instance, the tidy traveler from the Missouri River who had packed his slabs of bacon atop the batteries of his mill and later wondered why the gold ore worked by this grease-caked machine did not amalgamate with quicksilver. To prevent further serio-comic episodes of the inexperienced dealing with the "nicest principles" of mechanics and metallurgy, two engineers, Silas W. Burt and Edward L. Berthoud, suggested closer attention to mill operations: careful checks

[5] Frank Fossett, *Colorado, Its Gold and Silver Mines, Farms and Stock Ranges, and Health and Pleasure Resorts*, pp. 132, 266; Hollister, *Mines of Colorado*, pp. 112–13; Pullman to family, September 17, 28, and October 31, 1860, Pullman Papers.

on the ore to determine its quality; better maintenance of machinery; uniform flow of the ore onto the batteries; and filtering the mercury, nine-tenths of which, they claimed, reached the territory impure. They noted as well that the freshly cut pine wood used in fabricating the mills tended to shrink, thus opening the joints and causing the whole structure to shake itself apart.[6]

For miners bent on their own solutions, there were experiments in exposing the stubborn sulphurets to the air a few days in hopes of instant oxidization, or boiling them in vats of mercury, or, of sounder principle, crushing them under heavier stamps and at higher rates of speed. None of these schemes seemed to work, and returns dwindled to the point where further development of the hardrock mines was in jeopardy. As practical miners fully understood, the relationship between mine and mill was crucial; unless the ore could be converted into gold, there would be no way to pay the costs of tedious hand drilling, blasting, timbering the shafts, and hoisting the ore, which resulted from following the veins deeper and deeper into the mountain sides. Windlasses cranked by hand or whims harnessed to horses had been sufficient to bring up ore, waste, and water from shallower pits but proved woefully inadequate in shafts sunk beyond two hundred feet. Steam-powered hoists and pumps became necessary, along with the black powder for blasting away the rock. Essential, too, for purposes of shafting and as a fuel supply was timber, which rising demand and foolishly wasteful practices had been making ever harder and more expensive to obtain.[7]

Aggravated by the outbreak of the Civil War, a general despondency settled over Colorado mining and milling in 1861 and 1862. The days when a poor man with grit and a strong back could make surface deposits pay were fleeting fast, and the future rested precariously on deep mines and inefficient mills. Perhaps with some foresight, John H. Gregory had sold his two claims in the summer of 1859 for $21,000 (to be paid in time from the product of the mines), returned to prospecting, and hired out his services as a mining expert, reputedly earning two hundred dollars a day for his advice. Until 1862, a series of lessees, including Pullman, worked por-

---

[6] Jerome C. Smiley, *History of Denver*, p. 393; S. W. Burt and E. L. Berthoud, *The Rocky Mountain Gold Regions*, pp. 44–50.
[7] Fossett, *Gold and Silver Mines*, p. 205; *MSP* 6 (April 20, 1863): 2.

tions of the property with minimal success and, being unable to pay expenses from the refractory ore and the hardships of lode mining, returned the claims to their owners.[8]

The mines had simply moved beyond both the means and the ability of their operators to push extensive development, and hopes proved illusory that local capital could do the job. Instead, capital was a scarce commodity in the territory, and, when it could be found, it often shied away from the hazards of mining or commanded steep rates of interest, usually a minimum of 5 percent per month. As revealed in his letters home, Pullman, at first fascinated by the prospect of mining in the Rockies, increasingly invested his "fair" amount of capital and business skill in nonmining enterprises; to his mining and milling interests, he quickly added a lumber mill, a ranch at Golden, a freighting outfit, a hay farm, a store, and an active trade in gold dust. Personally aware of the risks, he reportedly loaned money to miners at monthly interest rates of 25 to 50 percent. Moreover, neither the placer mines nor the early stages of quartz mining had produced sums of money in reserve that might now have been employed in deep-level work. Thus, revival and expansion of the mining industry awaited outside investment.[9]

From its inception, mining in Colorado had been mainly the result of eastern interest and initiative. While geography and the extraordinary richness of Nevada's Comstock Lode kept Californians mostly west of the Rockies, the news of gold at Pike's Peak had drifted through dozens of towns and villages along the Missouri, Mississippi, and Ohio rivers and middle westerners had taken the lead in dashing for the mountains. Rarely did the migration to a new economic frontier result in an immediate or total break with native states and regions, and, at times of need, whether for money, machinery, or consolation, the pioneers in Colorado could be found writing home. Letters containing personal adventures or descriptions of the mines generated interest, if not always a burning desire to drop all and head for the West, among friends, family, and neigh-

[8] Hollister, *Mines of Colorado*, pp. 63–65; Thomas A. Rickard, *The Romance of Mining*, p. 216.

[9] Pullman to family, August 20, September 17, October 9 and 31, November 25, December 2, 1860, January 28, 1861, Pullman Papers; *Denver Post*, December 31, 1903.

bors back in "America." Small bankers and merchants, usually on the Missouri River frontier, sensed the profit potential of the rush and dispatched representatives to Colorado to set up branch houses through which supplies and services of all kinds could be sold for gold dust. Faulty and unreliable though these early contacts were, they nonetheless formed a chain of communication and trade that linked West with East and undeveloped areas with financial and manufacturing centers; this was a link that had the capacity to grow. David H. Moffat, Jr., for instance, who had come to Denver in 1860 as the agent and partner of a stationer in St. Joseph, Missouri, enlarged his original business of selling books, magazines, newspapers, and even wall paper to the miners into a small store that sold food staples from the Middle West which were in short supply and brought high prices. Eventually, by associating with an Omaha bank, he bought and sold gold dust, loaned money, and conducted a private banking business. By 1863 several banking houses in Omaha, Atchison, and Leavenworth, motivated chiefly by the abundance of gold dust, had established branches in Colorado to purchase the metal, receive deposits, issue notes, handle collections, and make loans on a personal basis at interest rates of 5 to 20 percent.[10]

In this exchange, substantial sums of gold dust passed into the hands of pioneer merchants and bankers and began the journey eastward through the opening channels of trade and finance. At times as much as forty thousand dollars worth of Colorado gold departed Denver on a single stagecoach headed for towns on the Missouri River, and there were always smaller amounts carried "back home" by friends of those miners who objected to the high gold-handling fees charged by the stage lines and express companies. From the inland towns the metal moved further east to the financial and manufacturing centers of the Middle West and the Northeast, with New York City being the ultimate destination for much of it. One source reported that New York received over $8 million in gold from Colorado during the year ending August, 1863, and, although a gross overstatement, it indicated that the flow of precious metal from the

---

[10] Ezra Millard to Moffat, March 2, 1861; Lee, Ripley and Company to Moffat, April 22, 1861; Joseph H. Millard to Moffat, April 27, 1861; all in David H. Moffat, Jr., Papers, First National Bank of Denver Archives. Fred R. Niehaus, *Development of Banking in Colorado*, pp. 14–16.

Rockies had reached major proportions, and at a moment when receipts of California gold had fallen off. No doubt these shipments would have occasioned the interest and excitement of a gold-conscious people at any time, but in the convulsive and uncertain climate of wartime the reaction of New York was swift.[11]

Serving to intensify eastern interest in the Colorado mines was the impact on the nation of two economic policies adopted by the government within the first twelve months of the war: the suspension of specie payments at the close of 1861 and the issuance of legal tender paper money, almost at once called "greenbacks" by the public, in early 1862. Designed to ease the financial drain on the Union, these measures made gold a more valuable and desirable commodity and an object of speculation. The greenbacks, which totaled $450 million by the end of the war, fostered speculation by their fiduciary character, so that every political or military crisis that affected public confidence in the government was reflected in the value of this unsecured paper money. Accordingly, the sagging fortunes of war created desires to exchange the greenbacks for gold, stocks, or other items considered to be more valuable. To paper money as well can be traced the growing sense of prosperity that seized the public mind and made more readily acceptable the risky opportunities of mining. And, finally, for businessmen, always a potent force in setting the tone and direction of the larger economic community, rising profits amid fluctuating prices apparently created visions of still greater wealth through timely investments in the mines.[12]

These forces unleashed by the war opened the way for a multitude of speculative ventures in the East that ranged from gold and railroads to petroleum, whisky, and mining. Gaining strength in the late summer of 1863, the urge to trade inflated greenbacks for securities drew throngs of "operators" into the financial district of New York, where they shoved into brokers' offices or jostled each other on street corners to catch the words of curbstone brokers who bought and sold in the open air. Carriages reportedly choked lower Broad-

11 Colorado's entire gold production for the period 1858–1867 was $25,021,784 according to Charles W. Henderson, *Mining in Colorado: A History of Discovery, Development and Production*, U.S. Geological Survey Professional Paper 138, p. 88. *MSP* 7 (August 3, 1863): 4; Wesley C. Mitchell, *A History of the Greenbacks*, p. 192.

12 Mitchell, *Greenbacks*, pp. 395–97.

way, and messenger boys darted adroitly from building to building
with the latest transactions. Although the tempo on Wall Street varied
with the stock offerings and the news from the military and political
fronts, between that summer and the spring of 1864 a "speculative
frenzy" gripped the city's stock market and was repeated on Chestnut
Street in Philadelphia and State Street in Boston. One veteran of Wall
Street booms recalled that all previous speculations had been mere
"ripples on the ocean of finance" compared with the "great tide
wave . . . rather an earthquake wave of the tropics" that swept through
his domain.[13]

From the standpoint of the shaky new enterprises on the Rocky
Mountain frontier, being scarcely known or understood and badly in
need of capital for development, it might have seemed that no better
economic conditions could prevail in the East than those of 1863 and
1864. The initial work on the gold lodes had shown the area's potential worth, but within sight now were the elements essential to
building a permanent and prosperous industry: outside interest and
financial support, with the promise of mechanization, professional
management, and engineering and scientific skill. However, whether
there would be a mutually beneficial fusion of western resources and
eastern know-how depended upon the attitudes of both buyers and
sellers. The same speculative fever that would produce a rich market
for Colorado mining properties was fraught with the danger that
overanxious western mine owners might underplay the practical
problems of deep-level mining and that enthusiastic eastern purchasers might invest foolishly. Of course, this possibility would be
heightened by greed, inexperience, distance, wartime, and similar
complicating factors.

In these hectic days, gold mining captured the imagination of
New York businessmen. Once-conservative merchants and bankers,
shedding suspicions born in tighter times about the wisdom of mining
proposals, ushered Colorado representatives into their offices and
listened in rapt attention to tales of gold deposits that were, oddly,

---

[13] For a general treatment of speculation in mining and petroleum during the
war, see Emerson D. Fite, *Social and Industrial Conditions in the North During the
Civil War*, pp. 24–34. James K. Medbury, *Men and Mysteries of Wall Street*, pp.
8–10, 277–79; William W. Fowler, *Twenty Years of Inside Life in Wall Street*, p.
298.

poor but rich, situated far off in the wilderness yet close to home. Said one bemused witness, there was news of "hidden treasures without number, materials for wealth without stint, ungarnered, unminted, unmoulded, almost unwatched." It was common for the banker or merchant, if convinced by such reports, to propose that a stock company be formed on the basis of the mining property, the owner to be paid from the first shares of stock sold, a board of trustees and company officers to be selected, and the enterprise to be chartered under the general mining and manufacturing laws of the state. The company thus organized and duly registered, next came the printing of stock certificates, the preparation of advertisements, and the opening of a subscription list to investors, who were urged to purchase the limited number of shares being made available to the public before prices soared beyond the reach of all but the very wealthy. With few alterations, this was the method adopted time and again during the Civil War to introduce mining ventures on eastern money markets.[14]

An early entrant from Colorado was the Ophir Mining Company, organized in October, 1863. All signs indicate that a group of New Yorkers paid nearly $200,000 for approximately 460 feet of gold-bearing ground in the Gregory District that had belonged to an eccentric Irishman, Pat D. Casey, who had struck it rich after a string of misfortunes. Itching to impress friends with his newly won wealth, Casey supposedly bought a span of black horses for $500 but grandly proclaimed them worth $2,500 and dutifully paid taxes on the greater amount. However, he came closer to real prosperity when he placed his mining properties in the hands of a trio of local promoters —including the organizer of the Pony Express, William H. Russell— for sale in the East and received about $60,000 for himself and enough to pay each of his agents a handsome commission. While others began the scramble for shares in the Ophir, Casey arrived in New York, purchased a luxurious carriage, hired attendants, and planned a new career in gold and stock speculation, which he apparently deemed safer than mining.[15]

Whatever the fate of Casey's personal plans, his presence in the

14 Fowler, *Wall Street*, pp. 300–02.

15 Frank Fossett, *Colorado: A Historical, Descriptive and Statistical Work on the Rocky Mountain Gold and Silver Mining Regions*, pp. 277–79; Edward Bliss, *A Brief History of the New Gold Regions of Colorado Territory*, p. 19.

East fed the excitement in the Colorado mines. Especially aware of this aspect were the numerous promoters who filtered in from the West, their pockets bulging with deeds to remote mining claims, and placed a high premium on being able to refer prospective investors to some tangible evidence of the territory's worth. Between 1863 and 1865, William H. Russell, for one, tied the "magic" of his name among New York businessmen with carefully arranged shipments of gold from the West to aid him in organizing some ten companies to operate in Gilpin and Clear Creek counties. Yet, when selling shares in the new mining companies required additional techniques, brokers and promoters proved resourceful. To create a mood of sure, almost unquestioned, integrity about the business of mining, they outfitted company offices with cabinets displaying ore specimens, always rich in metal, shelves here and there with models of mining operations and perhaps a rock formation moulded in clay, and maps of the gold regions. To give a sense of prosperity, there were Havana cigars and a bottle or two of aged whisky and, most importantly, to remind visitors that they too could share in this fortunate enterprise, there was certain to be a stack of prospectuses and the open subscription book close at hand. One broker further believed that a first-floor office was indispensible for the stock speculator who "dislikes to exercise his gastrocuemius [*sic*] by climbing stairs." Primarily because of the large number of people in the East who were totally ignorant of the science of mining, many took delight in those promoters who sprinkled their speech with such mining terms as "lodes," "ledges," "pyrites," "sulphurets," and "true fissure veins," or in the mineralogists and geologists, their sudden expertise passing unnoticed, who confidently advised clients to buy stock after dating and assaying a few fragments of rock.[16]

Publicity became a sine qua non in opening eastern purses for any new venture, and the Colorado promoter who failed to dip into these murky waters was not worthy of the title. Though each had his own taste for technique, one proven device was the prospectus, issued by the company to present its officers and directors, the nature and extent of its mining properties, and basic information on its capitalization, particularly the number and par value of its shares.

[16] *MSP* 8 (May 28, 1864): 369; Fossett, *Historical, Descriptive and Statistical Work*, pp. 88–89; Fowler, *Wall Street*, pp. 303–04.

The prospectus also included a statement of the enterprise's splendid chances for paying dividends. Most of the nearly two hundred companies formed in the East to work in Gilpin, Clear Creek, and Boulder counties prepared one of these documents and strategically placed copies in the offices of mining brokers, sent them to known mining investors, and, in some cases, even distributed them on street corners and in the doorways of the stock exchanges. Simultaneously, advertisements were inserted in the daily and weekly journals, and press releases were furnished to the newspapers for their financial columns. Usually the message was deceptively simple: here was an unrivaled opportunity to enrich oneself in a western territory that lacked only money and aggressiveness to unlock its golden vault. Although promoters artfully combined the desire of Colorado to see its mines developed and the eastern enthusiasm for speculation, they tended to overstress what capital could do. The Standard Gold Company reprinted the testimonial of the commissioner of deeds for Colorado Territory that "all Clear Creek County has needed is capital, and when capital has been used you cannot find a single instance when the parties so using it have failed." With money from the eastern states, puffed another prospectus, Colorado would soon easily outstrip California and Nevada in the production of precious metals.[17]

In the meantime the greatest activity in mining occurred on the stock market, where brokers worked overtime in the winter and spring of 1864 to satisfy the mounting demand for mining stocks. A long line of eager investors stretched from office to sidewalk when the Consolidated Gregory Gold Mining Company, the discovery claims of Gregory forming a portion of its property, threw open its subscription books at the end of March and, of interest to an old hand on Wall Street, attracted a number of "outsiders"—an uptown milliner with four crushed one thousand dollar bills in his hand, a country minister wanting fifty shares, and several doctors, lawyers, and mechanics. In one hour this crowd reportedly bought fifty thousand shares of Consolidated Gregory at fifty dollars each and resisted attempts by late comers to buy away a few thousand of those shares at a 2 percent premium. Elsewhere, after investors had pleaded with officers of the Gunnell Gold Mining Company, stock

[17] Standard Gold Company of Colorado, *Prospectus and Report*, p. 23; Kip and Buell Gold Company, *Prospectus*, p. 6.

said to be reserved for the promoters was magnanimously put on sale and the company realized $100,000 more than its original capitalization. At other times, the frenzy fatigued buyer and seller alike: tempers flared within the ranks of impatient investors, occasionally leading to shoving and shouting matches that degenerated into fisticuffs, and one company, having sold $600,000 worth of its stock at par during the day, summoned a squad of police to clear out its offices at closing time. The sturdier speculator, however, could continue his pursuit far into the night at an evening exchange that opened uptown in the spring of 1864, or in the hallways and reading rooms of the Fifth Avenue Hotel, the popular gathering place for dealers in all kinds of securities.[18]

The boom psychology afforded a perfect opportunity for swindlers, East and West, to take advantage of the gullible and the impetuous who believed that Colorado offered so much for so little. In the mountains where rickety stamp mills hammered ineffectively on tons of refractory ore, word that millions of dollars awaited western mining schemes proved too great a temptation, reported one early historian, for "thousands," including many formerly content with the slow progress of the mines, who left their jobs and began to gather up mining properties for sale in the East. "The personal business of every man was to sell what he had, or possessing nothing, to hunt up a hole that might be put on the market," said another. Thus, under the guise of guaranteed bonanzas, hundreds of worthless prospects flooded the East to be purchased by anxious speculators. Among the most unfortunate would be the owners of "extension" properties—meaning the unexplored land that adjoined a known gold deposit but which was sometimes miles away and totally barren—or the investors in companies based on mine fragments, a few feet on several different lodes, which would be entirely unmanageable. Big losers in the latter category was a group of Roxbury, Massachusetts, businessmen who paid $60,000 to an ex-minister from Chicago for parts of 856 claims scattered all over Gilpin County. Coinciding with these

---

[18] The *New York Journal of Commerce* noted that it was the "minister, the doctor, the lawyer, the merchant, the shipowner, the banker, and . . . the newspaper writer himself who took the lead in mining investments" (in *MSP* 8 [May 28, 1864]: 369). Fowler, *Wall Street*, pp. 310–11; Medbury, *Men and Mysteries*, p. 253; *AJM* 1 (March 31, 1866): 5.

hoaxes imported from the West were the frauds, manipulations, and misrepresentations of eastern operators who swindled stockholders and, in a few instances, had the "cheek" to organize mining companies without the slightest intention of buying mineral ground, leading a federal commissioner to report in 1870 that there had been "a great many Hamlets with Hamlet left out" in Colorado.[19]

This brief but intense rage for Colorado mining ventures, tainted as it was by gross misconduct, obscured for contemporaries and historians alike the actual sum total of capital invested by easterners in the 1860s. In the absence of accurate data, witnesses to the excitement drew upon diverse and often untrustworthy sources to reach estimates that varied greatly as to the amount of capital, reflected nominal capitalizations, or covered different time spans. Near the peak of the boom the *New York Herald* reported that New Yorkers and Bostonians had invested over $150 million in Colorado mining companies. A Pacific Coast trade journal, depending on territorial newspapers, later said that $40 million had been spent over fifteen months. Almarin B. Paul, an experienced California miner, having traveled several months in the Rockies, claimed in 1870 that Boston capitalists alone had pumped $19 million into the heart of Gilpin County's gold district at Central City, Black Hawk, and Nevadaville. On the low side was an estimate of $8 million (nearly half of which went for the purchase of mining ground) spent over a ten-year period; this amount was reached by a Colorado mining historian in 1880 after "careful investigation." But the figure most often quoted was $20 million, and, from the information available, this seems to be the minimal extent of eastern financial involvement to the mid-1860s. Since eastern investors entered mining with more spirit than knowledge and more capital than caution, they paid far too much for the untested Colorado mines and compounded that error by subsequently overcapitalizing their companies and inviting serious financial trouble. They splurged valuable operating capital on the purchase of "feet" along gold lodes, paying as much as $1,000 a foot on the Gregory or $300,000 for 480 feet on less promising land. Quite commonly this extravagance resulted in the overcapitalization of an enterprise and

[19] Hollister, *Mines of Colorado*, pp. 207–08; *AJM* 2 (November 17, 1866): 120; Rossiter W. Raymond, *Statistics of Mines and Mining in the States and Territories West of the Rocky Mountains* (1870), p. 349.

the appearance of million-dollar companies on the eastern market. One account of the period 1860–1866 discovered that forty-five companies organized in New York had a combined nominal capitalization of $78.5 million and obviously bore little relation to what the territorial mines could actually produce at that time.[20]

Removed from the glitter of company prospectuses and zealous promoters, the challenge of developing the Colorado mines into dividend-paying propositions awaited the new eastern owners. In a sense, the chore was to build an industrial province on the raw frontier. If indeed the mining craze were to mature into a sound and permanent industry and not remain simply a colorful tile in the mosaic of speculation, then the dollars and dreams of these absentee companies would have to be transformed into rational business organizations that proceeded carefully and deliberately. Such was not to be the case. To the consternation of many, events would show the loss of huge sums of money to the bugbears of waste, mismanagement, and ignorance, and the territory, once believing eastern capital held the key to prosperity, would tumble into a deep decline.

An early problem was finding a manager capable of supervising the daily development of the company's property, a task that demanded men who, besides enduring the rugged environment, could direct the multiple operations of mine and mill, from drilling and blasting to reducing the ore. Because the job carried heavy responsibilities, some companies retained the discoverer or the promoter and trusted to his knowledge of the mine. But others, either of necessity, in the naive belief that mining was merely digging golden metal from the earth, or in the conviction that only a manager picked from within the company could be sure of displaying the "judgment, frugality, and integrity" needed for the job, chose easterners and novices. Consequently mine owners sent friends, relatives, and business associates, sometimes those down on their luck, into the Rockies where their inexperience, loose spending, and general lack of common sense helped make the mines greater consumers of money than producers of wealth. Attired in corduroy suits and polished brogans, there were some among these eastern agents whose main interest was

[20] *MSP* 8 (June 11, 1864): 412; *MSP* 10 (April 1, 1865): 199; *MSP* 20 (April 16, 1870): 250; *New York Times*, September 16, 1874, p. 9; Fossett, *Gold and Silver Mines*, p. 136; *American Mining Index* 3 (February 5, 1866): 8.

in "riding fast horses or tyrannizing their employees," and several "jolly-dogs" who were "excellent masters of the billiard cue, with uncommon pride in high boots and spurs, whose champagne bills were charged to 'candles,' and whose costly incense to Venus appeared on the books as 'cash paid for mercury.' " While some drew salaries as high as $10,000 to $15,000 a year, a Central City resident grumbled that their blunders were so great that they should "thank God, with the doctors, that the ground covers their bad work."[21]

To a degree the misfortunes that plagued General Fitz-John Porter as agent for the Gunnell Gold Mining Company in Black Hawk were typical. Porter, a graduate of West Point, came to Colorado in 1864, after having been court-martialed and cashiered for failing to obey orders at the Second Battle of Bull Run, and at a salary of $15,000 a year undertook the direction of all facets of mining for the company, as well as the construction of a mill. In short order, legal tangles, disputes over neighboring claims, high costs, and extravagance combined to shorten the life of the enterprise. Certainly Porter made his most costly error in constructing an elaborate stone building for milling machinery some three miles away from his mines and then suffering a final setback when the machinery ordered from the East had to be sold to pay its freight. The structure, dubbed "Fitz John's Folly," cost the company $150,000, a "large sum," said Bayard Taylor in an uncharitable reference to Porter's court-martial, in order to "repeat the experience of the national government."[22]

Time and again mill building appeared as a leading cause of failure in the 1860s. The leap-before-you-look attitude that often characterized American industrial expansion in the last half of the nineteenth century described the headlong rush by the first companies in Colorado to erect expensive reduction works before developing their mines and determining the size and quality of the ore bodies. In 1867 Ovando Hollister counted 1,700 stamps (the pestles for crushing ore) in ninety-five mills, and some of these mills had a capacity of fifty to one hundred tons of ore a day, with boilers weigh-

---

[21] John Wetherbee, Jr., *A Brief Sketch of Colorado Territory and the Gold Mines of that Region*, p. 19; Kemp, *Colorado's Little Kingdom*, p. 121; *MSP* 16 (February 1, 1868): 70; *AJM* 2 (September 15, 1866): 386.

[22] *DAB*, s.v. "Porter, Fitz-John"; *New York Times*, February 25, 1866, p. 6; Bayard Taylor, *Colorado: A Summer Trip*, p. 56.

ing up to 1,500 pounds and steam engines developing one hundred and fifty horsepower, which, as he noted, were "extravagantly disproportioned to the mine they were intended for." A few were so "elegant" as to be "more fit for boudoirs or saloons than for the purpose of mining." And, most importantly, nearly all of this heavy machinery had come hundreds of miles across the plains from eastern foundries.[23]

In the summer of 1864 a transfer agent at Atchison, Missouri, reported over 1 million pounds of mining machinery consigned to him for shipment to Colorado and estimated that it would take at least 1,000 wagons drawn by 12,000 oxen to move this mountain of iron to the Rockies. Similarly, the middle border towns of Omaha, Leavenworth, and St. Joseph enjoyed a boom from the traffic in mining paraphernalia and supplies. With completion of the transcontinental railway still several years away, the equipment had to be strapped to wagons and nursed over hundreds of miles of rough, inhospitable country for six to eight weeks before reaching Denver. Along the way the lumbering wagons were harassed by the raiders of Confederate general Sterling Price in western Missouri and on the plains by hostile bands of Sioux, Cheyennes, and Arapahoes who took advantage of troop withdrawals for the Civil War to step up their attacks on whites. As a consequence, high freight rates prevailed, usually between 10 and 30 cents a pound from the Missouri River to Denver, and one source placed the average cost of sending a pound of cargo from Leavenworth to Colorado at 15.43 cents in 1865. Multiplied over tons of machinery and other gear, this expense readily strained a new mining company's financial resources and sapped the strength of the entire undertaking. For example, the Black Hawk Mining Company, after exhausting its entire working capital to purchase a one-hundred-stamp mill, paid 19.5 cents a pound to ship the 200,000 pounds of machinery from St. Louis to Gilpin County.[24]

[23] The *AJM* estimated that by 1866 there were forty or fifty companies that had spent the full sum of their working capital on stamp mills and thus could not afford to develop their mining ground (2 [October 27, 1866]: 74). Hollister, *Mines of Colorado*, pp. 137, 432–33; J. Parker Whitney, *Colorado, in the United States of America*, p. 57.

[24] *MSP* 9 (August 6, 1864): 86; *MSP* 23 (August 12, 1871): 83; John Bennett, "Mining and Smelting in Colorado," copy, Colorado State Historical Society,

In spite of the enormous expenses involved in buying, freighting, and installing these huge stamp mills, eastern companies simply repeated the unhappy experience of the pioneers on a grander scale. It became clear that the sulphurets taken from the deep mines resisted the heavier machinery in the mills and that a greater part of the precious metal still escaped amalgamation—a bare one-third to one-half of the gold content being saved, while the remainder ran from the batteries into nearby gulches. Under these conditions, intelligent mining men in the East and West began to search for a scientific process that would remove the sulphur from the ore and free the gold for amalgamation, thus increasing the yield of gold per ton of ore. If the *New York Times* of December, 1864, was correct in thinking that the stubborn ores of Colorado demanded "more associative and less individual" mining than that of California and that the corporate invasion from the East was "probably fortunate for the future of this important territory," then the time to commit potent new devices and processes based on chemistry and metallurgy had arrived.[25]

The immediate result, however, was a process mania that swept over the Colorado mines between 1864 and 1867 and spawned a bewildering number of new schemes and variations on old ones to remove or decompose the sulphur in the ore. So many "experts" in ore reduction flocked to the territory that Coloradans began addressing anyone who displayed the least bit of knowledge about mining as "professor" and, in the ailing condition of their mines, unwittingly welcomed an army of metallurgical hierophants who were usually either outright charlatans or inexperienced, impractical, and overweening visionaries. The typical rogue, went one description, was a "Professor Toothorn" whose method included an "improved-back-action-lightening-gum-elastic-cylinder-and-Spanish-fly amalgamator, with which he can draw gold from a Rocky Mountain turnip." Actual examples were plentiful. In the summer of 1865, Dr. James C. Ayer, a patent-medicine king renowned for his popular curealls of Cherry Pectoral and Cathartic Pills, after reputedly conducting a series of

Denver; *American Mining Gazette, and Geological Magazine* 2 (August 1865): 468–72.

25 *New York Times*, December 12, 1864, p. 4.

tests at his Lowell, Massachusetts, home, introduced a saline and alkaline bath that "disintegrates the rock into its atoms, while it does not divide the precious metals therein contained. The extraordinary molecular separation of its particles, is shown by the fact that the hardest quartz, although as clear as glass, will, after being subjected to this process, absorb water like a sponge." Shortly thereafter the Chemical Gold and Silver Reducing Company, headed by soldier-politician Benjamin F. Butler, formed in New York to make Ayer's process available to Colorado, along with the promise that it would save 80 to 95 percent of the precious metal and cost no more than conventional stamp milling. To "palm off" his process, said one mining paper, Ayer got the weak endorsements of Professor Benjamin Silliman, Jr., of Yale College and United States assayer John Torrey that the treatment worked in the laboratory. But Ayer's nostrum failed in practical application, as did the process of New York chemist N. Kent, who displayed a culinary bent when he recommended that the ore be ground finely, mixed with a little salt and water to form a dough, molded into loaves, and baked in a kiln until the sulphurets decomposed. Another failure was the machine at Empire that looked like a "big ice cream freezer," belched clouds of steam, and used two hundred dollars worth of mercury every fifteen minutes.[26]

As most of these processes originated in eastern laboratories, what testing might have been done was limited in scope and time. Nor did the federal government act on the suggestion of a mining journal that an agency be created to perfect a means of treating the Colorado sulphurets. Under these circumstances, overeager mining companies selected a process on little more than the promoter's word and undertook the very great risk and expense of putting an untried technique into operation on the frontier, becoming to a degree the pawn as well as the patron of an experiment in metallurgy. Not surprisingly, the results were almost always disastrous for eastern com-

[26] Along with Ayer and Butler, the board of trustees of the process company included former governor of Utah Territory Frank Fuller, former governor of Massachusetts and Congressman George S. Boutwell, and U.S. Senator James W. Nye of Nevada (*Chemical Gold and Silver Ore Reducing Company*, p. 1). Charles Cowley, *Reminiscences of James C. Ayer and the Town of Ayer*, pp. 59–61; Samuel S. Wallihan and T. O. Bigney, *The Rocky Mountain Directory and Colorado Gazetteer for 1871*, p. 245; Fossett, *Historical, Descriptive and Statistical Work*, p. 93; *MSP* 10 (June 10, 1865): 360; *MSP* 11 (July 22, 1865): 40.

panies. "Probably the most ruinous," thought Rossiter Raymond, was the Crosby and Thompson process which, meant to vaporize gold as though it were a gas, failed to work for about thirty companies. Silliman, conveniently ignoring his connection with Ayer's scheme, declared Colorado "cursed" by foolish processes. He cited the example of the Consolidated Gregory Company, which spent $2.5 million on a technique to remove nonexistent lead from its ore, forcing the Yale scientist to conclude that its furnace was not "worth the bricks used in its construction." Thus, with their treasuries nearly depleted, companies painfully watched the repeated failures of their "process men."

Rich ores, from an adjacent mine, are on hand; steam is raised; the shrill shriek of the whistle affrights the mountain sheep in the ravines, and startles the prospector on the mountains; cylinders revolve; ball pulverizers clatter; red flames and blue shoot out from the mouths of heated furnaces; great volumes of smoke and sulpherous acid fumes go up the towering chimney, and gold and silver, too, for all the professor knows—at least, he never finds any of it worth mentioning— and then another failure.[27]

Severely damaged by the process mania along with high costs of every kind, poor management, and frenzied finance, some companies passed into receivership and eventually were dissolved, others simply suspended operations, and still others gradually returned to the inefficient practices of stamp milling and amalgamation. By 1866–1867 tons of machinery and assorted contraptions imported from the East lay idle, creating in Colorado a "scene of exorbitant hopes and equally extravagant disappointments" to the eye of traveler Bayard Taylor in the summer of 1866. According to a reporter for *Harper's Monthly Magazine*, most reduction works around Central City were "as silent as the tomb."

Scattered in all directions around half-finished, roofless buildings can be seen boilers and engines, stamps and crushers, pans and amalgamators, and machinery of every kind, half buried in the soil, rusting and wasting, lying in the roads, even driven over by the traveler as he passes the wreck—a monument of one kind of Eastern mining.

[27] *AJM* 6 (October 3, 1868): 212; *MSP* 10 (May 20, 1865, and June 10, 1865): 310, 360; *MSP* 17 (December 5, 1868); 366; Wallihan and Bigney, *Rocky Mountain Directory*, p. 247; Raymond, *Statistics of Mines* (1870), p. 356.

A disillusioned Boulder resident echoed the theme that eastern capital and companies had been more of a "curse than a benefit" to the territory, though he added somewhat disingenuously that a share of the blame properly belonged to those "great boobies" in Colorado who had sold claims to men whose mistakes slandered the name of the region.[28]

No doubt impatience, ignorance, reckless expenditures, and similar faults were responsible for the "mistaken system" adopted by eastern companies and in large measure accounted for this poor showing in the 1860s. Originating in a period of wild speculation, this first major movement of eastern money to Colorado miscarried because of its own excesses and because ambition was a poor substitute for ability in coping with the practical problems of hardrock mining. However, to the investors of New York, Boston, and Chicago the territory itself became an anomaly, since its mining business verged on collapse despite an abundance of precious metals and the influx of outside capital, and the mines, for whatever reason, seemed more suited to separating an easterner from his greenbacks than to sending forth a golden treasure. It was, thought a western miner, a "lesson which eastern capitalists will remember with a twinge of the pocket nerve" for many years. Substantially agreeing with all of these explanations of what had gone wrong, mining engineer Raymond added in 1869 that "the scientific men without practice and the practical men without science, the honest men without capacity, and the smart men without honesty" had contributed to the depression.[29]

Although virtually every analysis of the hard times in Colorado condemned eastern practices, it should be remembered that eastern investors tried to establish an industry under the most unfavorable conditions of time and place. That they generally met defeat in this self-appointed task is hardly surprising, but the failure need not lessen or obscure the efforts made by eastern companies to free the mines from financially undernourished parties of prospectors and to

[28] Bayard Taylor, *Summer Trip*, p. 61; A. W. Hoyt, "Over the Plains to Colorado," *Harper's Monthly Magazine* 35 (June, 1867): 9; *AJM* 3 (May 4, 1867): 107; Samuel Bowles, *Our New West: Records of Travel Between the Mississippi River and the Pacific Ocean*, p. 182.

[29] *AJM* 1 (June 16, 1866, and July 14, 1866): 180, 242; Rossiter W. Raymond, *Mineral Resources of the States and Territories West of the Rocky Mountains*, p. 5; *EMJ* 19 (January 9, 1875): 21–22; *EMJ* 23 (May 12, 1877): 320.

introduce heavier machinery, sizable working forces, and some scientific approaches to the business of mining and ore treatment. The presence of eastern money was also responsible for the opening of many new mines and for the further exploration of older ones at depths not possible with the simpler tools and equipment of the pioneers. And in paying the bill for the disastrous process mania, easterners supported trial and error procedures on a frontier of knowledge that provided additional information on the nature and content of Colorado ores and acquainted scientists with the metallurgical needs of the territory. Moreover, as suggested by some contemporaries, most notably special commissioner for the United States Treasury Department James W. Taylor, the eastern adventure in the Rockies had demonstrated the wisdom of separating mining per se from ore reduction, so that specialized research could advance in each branch of the industry.[30]

Dramatic evidence that eastern finance could have a constructive role to play in the mining development of Colorado appeared in the fall of 1867 when Professor Nathaniel P. Hill, a chemist from Brown University, erected the Boston and Colorado Smelter at Black Hawk, an event which historian Jerome Smiley said "marked an epoch in the economic history of Colorado." With the financial backing of a group of Boston businessmen, Hill had been able to examine the refractory ores of Gilpin County in 1864 and subsequently to study milling and smelting techniques at Swansea, Wales, and Freiburg, Saxony, ancient and distinguished centers of mining and metallurgical science. Drawing heavily on European practices, Hill perfected a smelting process and designed a plant, overcoming numerous delays and setbacks along the way, and early in 1868 commenced operations by successfully treating higher grade ore from surrounding mines and shipping his product, a dull copper matte, to Swansea for refining. It was, of course, a personal triumph for Hill, eventually leading to a career in politics and a seat in the United States Senate. However, Boston capital had made possible much of the slow and tedious experimentation on the process and the construction of a smelter thousands of miles away at a time when each brick used in the furnaces cost one dollar, iron ran twenty-two cents a pound, and skilled laborers received eight

[30] James W. Taylor, *Report on the Mineral Resources of the United States East of the Rocky Mountains*, p. 345.

dollars a day and common laborers, four dollars. Furthermore, it is unlikely that this project of Hill and his Boston associates could have been undertaken at all without the economic incentives of eastern mining companies in search of a satisfactory process, and it certainly would not have come as soon as it did.[31]

Still, no one could dispute the fact that the first rush of eastern money to Colorado was on the whole a great debacle. And it was also clear that if Colorado were to rekindle eastern interest in its mines, it would need the assistance of countless promoters who could open or reopen the channels of investment by publicizing the untapped mineral resources of the state and by reviving the abused dream of the early 1860s that mining was the quickest way to a fortune.

[31] On the scientific accomplishments of the Boston and Colorado smelting works, see Rodman W. Paul, "Colorado as a Pioneer of Science in the Mining West," *Mississippi Valley Historical Review* 47 (June, 1960): 42–44. Smiley, *History of Denver*, p. 550; Thomas Tonge, "The Evolution of Mining and Ore Treatment in Colorado," *Engineering Magazine* 18 (1900): 269–71.

2

In Pursuit of Capital

An indomitable spirit of enterprise gripped the American nation in the last forty years of the nineteenth century. Whenever opportunity knocked, whether it be in clearing a virgin forest of its timber, opening the earth for its minerals, extending the railroad network, building a factory, or farming the latest agricultural Golconda, Americans responded to their "main chance" with amazing vitality. By the standards of the day, a Carnegie or Rockefeller, who recognized the rich potential of a growing country and chased it with single-minded devotion into the outer reaches of opulence, took on heroic dimensions and gave the age its distinctive entrepreneurial character. From one end of the United States to the other could be felt the throb of an ambitious, venturesome, optimistic people, who organized and schemed in imitation of the great financiers and hoped to share in the surging economic expansion of the nation through thousands of plans and projects to make a fortune.

To one degree or another every effort to gain public credence and capital investment for a new economic opportunity involved someone called the promoter. Although he served an important function in developing a business enterprise, the promoter labored under a tarnished reputation. In the seventeenth century Daniel Defoe identified promoters as men who "being master of more cunning than their neighbours, turn their thoughts to private methods of trick and cheat, a modern way of thieving . . . by which honest men are gulled with fair pretences to part from their money." That noisome image had not improved substantially in the America of the late nineteenth century. The promoter remained either a mysterious and treacherous character who lived on his wits and spent every spare moment in hatching elaborate schemes to victimize honest investors, or he was the less wicked but inept dreamer in the style of Colonel Beriah Sellers created by Mark Twain in *The Gilded Age*. In fact, while the suspicions were fully justified, the mining promoter, as well as his counterpart in other ventures, forged the essential links between financiers and projects that invigorated the economy; his self-confidence, his unflinching faith in the future, and his boosterism reflected prominent national characteristics. The sentiment behind the promoter's "boost, don't knock" slogan had deep roots and a familiar ring in a nation used to accentuating the positive and making the most of available opportunities.[1]

Since they were underdeveloped regions rich in natural resources, Colorado and its sister mining territories produced some of the liveliest promotion in the period after 1860. A recurring scene in the financial districts of major American cities, and also in many of those in Europe, was the visitor recently arrived from the West who scurried from business office to bank looking to enlist capital in his mining proposal. From New York a businessman reported in 1897 that this "town is full of Coloradans with mines to sell from every district of the State. I have an average of a report a day presented." And once in the mining districts themselves, the promoters became endemic. At the slightest hint that a traveler from the East represented capital, he was apt to be stopped in Colorado by a flock of local boosters—even in the most unlikely spots—wishing to unload a mining property. Writer Bayard Taylor was astonished in 1866 to find that not only did his hotel keeper have 25,000 feet on a gold lode for sale, but the desk clerk, the porter, and the chambermaid also had claims, and he pointed out that the druggist might as readily dispense "chlorides instead of asperiants." After days of hearing so much and so often about the rich mining ground beneath his feet, Taylor confessed to nearly asking a waiter to "put a little more sulphuret in my coffee." The moment he emerged from stagecoach or railroad car, an outsider would be surrounded by a circle of promoters who dogged his every step, and learning his name and home town from a cooperative employee in the hotel made it easier for them to catch his lapel on a dusty street corner and say a few words about a truly marvelous mining property or

---

[1] Daniel Defoe, "Of Projectors," in *The Earlier Life and Chief Earlier Works of Daniel Defoe*, p. 44.

to thrust a deed into his hands for consideration "at his leisure." It was strong tonic, and few men were known to leave Colorado without at least feeling the urge to take a "flier" in the mines.[2]

However, development of mineral resources in the area did not depend solely upon such impromptu contacts between tourists and local vendors, and it is well that it did not, for capital investment would have come slowly and irregularly. Instead, at most times there was a small army of determined promoters ready to search elsewhere for financial backing and willing to go to great lengths to engage prospective buyers. Among the most talented and energetic was Joel Parker Whitney, whose promotional activities on behalf of the Colorado mines and other western projects spanned forty years. A native New Englander, though raised in California not far from the gold fields, he had briefly followed a business career in Boston during the Civil War. At the end of the war, desiring, as he said, to feel the "exhilaration of a freedom I had long been denied," Whitney went to Colorado. A born adventurer and an avid sportsman, he may have explained his own behavior in leaving Boston for the Rockies by a parallel he saw between trout and mankind: "Why trout will remain about one place for life is difficult to explain; but they do. And so we may say about men. Why will they stay in one place and eke out an uncertain and precarious existence, when they can go where they could do so much better?" Nevertheless, the principal purpose of his trip in May, 1865, was to inspect the Colorado mines for several Boston businessmen—including Oakes Ames, a leading figure in the Crédit Mobilier Scandal—and, if a valuable property caught his eye, to invest $5,000 of the group's money. Subsequently, Whitney was chief organizer of the Bullion Consolidated Mining Company, incorporated in Massachusetts, which owned gold mines near Central City and silver properties in the Ten Mile District of Summit County. More importantly, the transactions began a lifetime in western promotion.[3]

As a "Bostonian mining operator," Whitney planned his projects

[2] William H. Brevoort to Eben Smith, February 24, 1897, Box 2, Eben Smith Papers, Denver Public Library; Bayard Taylor, *Colorado: A Summer Trip*, p. 59; Miguel A. Otero, *My Life on the Frontier, 1864–1882*, pp. 138–39.

[3] J. Parker Whitney, *Reminiscences of a Sportsman*, pp. 43, 110, 158; *Monterey* (Calif.) *Daily Cypress*, January 18, 1913; *Bullion Consolidated Mining Company*, p. 4; *American Mining Index* 3 (January 8, 1866): 15.

with great care and promoted them with a keen sense of the power of publicity. In the fall of 1865 he had hired several men from Central City to visit recently opened deposits in the vicinity and collect ore specimens, which were then assayed, labeled, and shipped to his office in Boston. At the same time Whitney apparently took options to purchase the better properties and agreed to raise development funds in the East or surrender the options after a stipulated period of time. Having made these arrangements in Colorado, the promoter hurried east and quickly produced a pamphlet entitled *Silver Mining Regions of Colorado* for the unabashed purpose of "attracting the attention of capitalists to the inviting fields" of the Rocky Mountains and identifying himself as a responsible dealer in mining properties. Except for a preposterous comparison between the exceedingly rich silver mines of ancient Mexico and those of Colorado based primarily on their equally high elevation, the booklet was on the whole a moderate piece of promotional literature that recognized the value of silver in Colorado and stressed the need for caution in entering the business of mining. Without mentioning Whitney's company specifically, the enthusiastic treatment of Summit County made it clear that there could be found the best and brightest prospects in the Rocky Mountains. Moreover, the pamphlet opened up greener pastures for the promotional activities of its author.[4]

Years later Whitney recalled that his early faith in the mineral resources of Colorado never wavered for a moment and that he continually sought new and better ways to advertise them. This appeared to be so in 1867 when he volunteered to serve as the delegate from the territory to the grand Paris Universal Exposition opening in the spring of that year. Owing to the depressed conditions of the mines and the lethargic response of mine owners, the first appointee, mining engineer George W. Maynard, found himself stranded in New York, along with several crates of ore samples, while friends in Colorado tried in vain to raise additional funds to finance his journey to France. When efforts to get support from the New York mining crowd also failed, Whitney, frankly admitting that his desires were "not purely eleemosynary," convinced Maynard to have the appointment trans-

---

[4] Whitney, *Reminiscences*, pp. 121, 127–28, 138, 173; J. Parker Whitney, *Silver Mining Regions of Colorado*, pp. 1, 13–41, 75–81; *MSP* 17 (August 22, 1868): 113.

ferred to him, thus becoming the territorial commissioner and agreeing to pay all expenses of installing an exhibit in Paris. Certain that
there were "hundreds of thousands who will never hear of our in-
ducements, unless we present them," the promoter departed for
Europe with his company prospectuses as well as his personal collection of Colorado ores.[5]

In his official capacity as commissioner from Colorado, Whitney performed his task well—enough so that a California journal
grouched that its state had been "overshadowed altogether by an
interior mining district, perched upon an isolated spur of the Rocky
Mountains." His handsome display of ore specimens, supposedly acquired from four to five hundred different lodes, won a gold medal
personally presented by Emperor Napoleon III, and a pamphlet of
his describing the chunks of ore and briefly outlining the history of
the territory appeared in English, French, and German editions and
circulated widely at the exposition; all of which was thought to have
had "salutory" effects upon foreign investors. As expected, he also
found time for some private promotional work and returned to Colorado with Louis L. Simonin, a mining engineer, professor of geology,
and agent for Napoleon III; a Monsieur Geise, a high official in the
powerful Crédit Foncier of Paris supposedly backed by sixty million
francs; and Colonel Wilhelm Heine, a wealthy German diplomat who
reportedly represented Frankfurt interests. If Whitney had hoped to
get European money to construct a smelter near his mining operations,
the project never materialized, but the three men did visit his mines
and were lavishly entertained by the superintendent of Bullion Consolidated.[6]

Through the 1890s, J. Parker Whitney affixed his name to a
succession of western business ventures, including mines at Leadville
and Cripple Creek, a Texas cattle ranch, copper mines and a railroad
in New Mexico, and even a school of agriculture to train wealthy
Englishmen to grow Sacramento Valley oranges, which qualified him

[5] *Denver Times*, February 2, 1901; Frank Hall, *History of the State of Colo-*
rado, 1: 440; *Central City* (Colo.) *Daily Miners' Register*, March 19, 1867; George
W. Maynard, "Early Colorado Days," *MSP* 98 (June 5, 1909): 791–92.

[6] The English version of Whitney's pamphlet was *Colorado, in the United*
States of America. Whitney, *Reminiscences*, p. 189; Hall, *History*, 1: 441; *MSP* 15
(October 12, 1867): 232; *AJM* 4 (September 28, 1867): 197; See also Louis L.
Simonin, *The Rocky Mountain West in 1867*, pp. 3, 44.

as a professional promoter. However, anyone who apprehended the economic potential of gold and silver in Colorado and the other western states and devised a plan to procure financial help could and frequently did become a mining promoter. Industrialists, merchants, lawyers, soldiers, physicians, politicians, men of the cloth, and people from equally diverse backgrounds filled the ranks of promoters in the post–Civil War decades. In the 1870s and 1880s, for instance, Chicago capitalists heard the virtues of mine investment extolled by a former hotel keeper in the city, an organizer of minstrel shows, and a professor of music and literature. And in the 1880s several northern cities, particularly New York and Chicago, followed the antics of newspaper publisher, politician, and humorist Marcus M. ("Brick") Pomeroy to finance a tunnel through the spine of the Rocky Mountains, by which he planned to connect Denver and Salt Lake City by rail and mine every deposit of precious metal he found from the bottom up.[7]

Predictably, there were times when the prospector himself, usually more accustomed to combing the hillsides and ravines for blossom rock than office buildings for financiers, bought some new duds, boarded the railroad or stagecoach in Denver, and tried his luck at selling a mine or two in the East. When the market for mining properties was strong and enthusiastic, as it was in the early 1860s, the discoverer of a mine stood a fair chance of placing his property in eastern hands, and in 1864 a perennial prospector named George H. Fryer returned to his native Philadelphia, where he reportedly made a "small fortune" from the sale of his Colorado claims. For the most part, however, before accepting the chore of promoting their own discoveries, prospectors would have benefited from the advice of a writer in 1882 who emphasized that perseverance was as important to selling a mine as to finding one. First, he instructed, jot down a history of your property in a notebook and gather up a few specimens before coming East to locate a buyer. Then, look up friends and relatives who may know a capitalist and try him at once, but, if that fails, walk "boldly" into a business office, disregarding any rebukes and denials, and see the man in charge. Finally, upon meeting a financier,

[7] *Boston Evening Transcript*, January 18, 1913; *EMJ* 65 (January 8, 1898): 48; Hall, *History*, 4: 150; Edgar C. McMechen, *The Moffat Tunnel of Colorado, An Epic of Empire*, 1: 69–70.

went these rules of conduct, do not idly "chin" or small talk about the weather or the trip from the mines, since "the morning paper has told him what the weather will be for the day and he never travels the emigrant train." Though this strategy offered no guarantee of success, the message was pointed: if you "prospect" for a capitalist, you will "locate" one.[8]

On occasion the professional mining engineer, ordinarily the impartial ally of cautious investors, also assumed the role of promoter to enlist outside capital. Because his job frequently took him to remote mining regions where he examined and reported on new mineral discoveries for absentee investors, the engineer could often do some exploring for himself and perhaps formulate a mining venture for friends and associates in the East. In 1891, John B. Farish, a highly respected engineer of wide experience in the West, attempted to organize a mining concern near Rico, in southwestern Colorado, an area then on the verge of a boom touched off by the sale of the Enterprise silver mine to a syndicate of New York and English capitalists for some two million dollars. Writing to Isaac L. Ellwood, an Illinois barbed-wire millionaire, Farish took credit for the sale of the Enterprise on which he reported, and he now invited Ellwood to join him in the "best and surest mining proposition I have ever seen for five years." His plan centered on developing the Aetna and Revenue groups of mines, which adjoined the Enterprise and covered approximately seventy-five acres of the choicest silver-bearing ground in the region. Although the engineer was "under the impression" that both groups could and should be purchased outright, he thought that Ellwood could bond and lease the Aetna for $75,000 and the Revenue for between $75,000 and $100,000, for one year in each case. Furthermore, since the properties had been opened by three well-constructed shafts, Farish estimated that $30,000 would pay for all development work, $12,000 of which would be needed at the outset to purchase and erect mining machinery. Thus, in presenting his scheme to the Illinois manufacturer, Farish reminded him that the Enterprise transaction "has turned the attention of speculators to the

8 Years later Fryer made the important New Discovery location at Leadville on what came to be called Fryer Hill (G. Thomas Ingham, *Digging Gold Among the Rockies*, p. 405). William R. Balch, comp., *The Mines, Miners and Mining Interests of the United States in 1882*, p. 853.

camp of Rico, and it will be necessary for prompt action to prevent prices [from] being run upon us," and he hoped that Ellwood would "proceed with some care and not let any one know the plans proposed herein except those who may join us." In return for his services as promoter, Farish asked one-fifth interest in the company, free of the purchase price and the expenses of development, and agreed to direct all operations at the mine once the transaction was completed.[9]

Due to what Farish termed the "outrageous figure" placed on one of the properties, he subsequently deemed the mining proposition at Rico "quite out of the question" and speedily withdrew from further negotiations. Undismayed, however, he contacted Ellwood with a new project based on a mine he had examined some years earlier and now proclaimed a "perfectly sure and safe mining enterprise," capable of producing $750,000 in profits over a three-year period. But there were complications in this endeavor, too, and, despite a persistent effort by the engineer to locate a suitable mine for Ellwood, several years passed before the two collaborated in the purchase of a few claims near Ouray and incorporated the Wedge Mines Company with Farish in the presidency and the barbed-wire king as principal owner.[10]

More often the highly trained mining engineer contentedly collected his fees as a consultant to mine operators or as a resident expert in the long-range development of a mineral property, rather than jeopardize his professional integrity or career in a mining promotion that might go awry. For different reasons, the locator of a claim in the Rockies did not usually travel outside the region to seek capital, partially because it took skill and money to plan and present a mining venture—money to visit eastern or European cities and money to track down financiers—which were beyond the means of the ordinary prospector. In the 1890s a promoter in New York stated he spent four

[9] Farish "attained high rank among the more eminent in his profession" (Hall, *History*, 4: 122). Farish to Ellwood, May 20, June 2, June 5, 1891, Box 18; Farish statement on the Newman Hill mines, Rico mining camp, Box 146; both in Isaac L. Ellwood Papers, Western History Research Center, University of Wyoming, Laramie.

[10] In the first six months of 1897, the Wedge Mines showed a net profit of $96,091.39 (Typed statement on Wedge Mines Company for 1897, Box 146, Elwood Papers). Farish to Ellwood, August 22, 1891, Box 18; October 9, 1894, December 31, 1894, Box 19; January 4, 1895, Box 27; April 4, 1895, Box 28; December 2, 1896, January 16, 1897, Box 37; all in Ellwood Papers.

thousand dollars in advertising one scheme for one year. Many years earlier the Central City correspondent of the *New York Times* noted that the people furthest from "civilization" were always the poorest ones and "none but those who have no possessions of any amount will leave their homes and occupations and go forth to try the chances of life in an unknown and far distant region." If, as the reporter observed, poor prospectors inhabited Colorado, months of searching after gold and silver exhausted whatever funds they may have had for food, clothing, and other necessities, and they could ill afford the costs of promotion. Furthermore, chances were slim that the frontier prospector possessed the techniques, social graces, ease, or influence to mix successfully in the monied circles of New York, Boston, or Chicago, where such enterprises were usually floated, or that he had friends or relatives of wealth sufficient to develop his discoveries. In practical terms, access to capital was not within the grasp of the average claim holder in the West.[11]

The variety of limitations on the engineer and the prospector, coupled with the remoteness of the Colorado mines, left the high stakes of mine promotion in more qualified and willing hands. Presumably, such middlemen had all of the friends, connections, and tact found wanting in the discoverer, and so they purchased or bonded his property and began the formation of a mining concern believing they would turn a fancy personal profit in the transaction. Exactly what sort of an arrangement would be made by the promoter for payment of his services depended on his own requirements, the wishes of the buyer, the strength of the enterprise, and related matters. Ordinarily the vendor took his fee in cash or, being a speculator, accepted shares of stock in the new company, which he could then sell or hold as he saw fit. J. Parker Whitney, who took his commissions in stock—as much as one-fourth to one-third of the total issue—confided to a friend that he earned $600,000 from sales and dividends on stocks he won in promoting three mining ventures. A New York vendor in 1895 sold two thousand shares in a Colorado operation and promptly received five hundred shares as his commission. A close student of the business revealed that 40 to 50 percent of the capital stock in some

11 *MSP* 55 (November 5, 1887): 292; L. Bradford Prince to J. J. Fitzgerrell, April 10, 1894, LeBaron Bradford Prince Papers, New Mexico State Records and Archives Center, Santa Fe; *New York Times*, September 9, 1866, p. 3.

companies went to the organizer, and he then roundly condemned this practice for siphoning off badly needed funds for development of the mines. In a different way the same point about the high fees paid mining promoters appeared in the story about the fellow who wondered why an excited crowd had gathered on a Leadville street corner, "Good God, don't you see?" he was asked, "There's Colonel . . . with his hand in his own pocket!"[12]

Actually, of course, placing a mine in the East was no easy matter, even for the most skillful salesman. Each stage in the transaction had its own problems and risks. Besides the initial chore of locating a property in the West and negotiating satisfactory terms with its owner, the promoter took responsibility for drawing in the chief financial backers (men of stature if he planned to sell stock to the public), which meant writing letters and making personal visits, perhaps arranging an examination of the mine when a prospective buyer insisted, as well as composing a prospectus, issuing broadsides and other advertisements, and doing everything else necessary to float the project on the market. In seeking capital for his Clear Creek Placer Company below Idaho Springs, the former governor of New Mexico Territory, L. Bradford Prince, reported to be a sharp operator, wrote no less than eighty-five "long" letters to likely investors. When the target of one of these missives was not known to him personally, Prince simply addressed it "Dear Sir," explaining that he had heard the fellow was a "gentleman who feels an interest in mining enterprises, and who has heretofore made some investments therein or contemplated doing so," and proceeded to offer him shares in a mine with "no element of chance or failure in it." "Not leaving a stone unturned," his letters fanned out across the country, from New York to Oklahoma, to small towns and major cities, and reached attorneys, clergymen, judges, businessmen, diplomats, a labor leader, and a member of Congress. Only one of these responded favorably, and then for a special price on a large block of shares.[13]

The trouble Prince experienced in catching small-fry investors

[12] Henry B. Clifford, *Rocks in the Road to Fortune or the Unsound Side of Mining*, p. 177; Whitney to Eben Smith, November 11, 1895, Box 1, Smith Papers; Seth G. Pope to Prince, August 4, 1895, Prince Papers; Jay F. Manning, *Leadville, Lake County and the Gold Belt*, p. 78.

[13] File folder on the Clear Creek Placer Company; Prince to J. J. Fitzgerrell, April 10, 1894; both in Prince Papers.

for his mining project might be multiplied many times over when a promoter set his sights on a first-rate businessman who would not be pushed into something until he examined it. A New York operator, J. N. Hayes, learned this in 1868 after inviting farm-machine manufacturer Cyrus H. McCormick, himself a superb salesman, to purchase "one-half of one of the most valuable mining properties that has been discovered on the continent," at a price of $40,000, plus $25,000 for working capital. Located near Georgetown, the property covered 6,400 feet on eight silver lodes that Hayes and his partners planned to exploit principally by driving a tunnel through majestic Sherman Mountain. To hurry the industrialist into an agreement, he was informed that the same opportunity would easily sell for $500,000 in England. But McCormick did not plunge—at least not this time. From mid-August to mid-October a steady stream of correspondence passed between Hayes, McCormick, and the financier's mining expert, Henry A. Ward, a professor of natural sciences at the University of Rochester. The letters mostly dealt with the accuracy of Hayes's description of the property and debated whether or not the promoters would pay for a mine inspection by Ward. At one point in the long bargaining, Hayes moaned that it was absolutely "insane to spend my time to negotiate a sale of it to Mr. McCormick." Indeed, matters soon became " 'all quiet on the Potomac,' " as Ward wrote McCormick, when the naturalist steadfastly refused to reduce his rate of $600 a month salary and $400 personal expenses, to be paid by the promoter, and rejected a one-sixteenth share in the mine in lieu of his fees for the examination. Turning elsewhere, Hayes eventually organized the Defiance Silver Mining Company with Civil War general Samuel P. Heintzelman in the presidency and apparently without the backing of Cyrus McCormick.[14]

As men of solid reputation and prominence in the United States, the McCormicks and Heintzelmans in all probability meant a great deal more to a designing promoter than simply their personal capital investment. If handled correctly, they might become magnets for

[14] Hayes to McCormick, August 15, 24, 28, September 1, 5, 12, 25, 1868; Hayes to Ward, September 12, 24, 1868; Ward to McCormick, October 3, 19, 1868; all in Series 2A, Box 34, Cyrus H. McCormick Papers, State Historical Society of Wisconsin, Madison. *DAB*, s.v. "Ward, Henry"; Defiance Silver Mining Company, *Prospectus and By-Laws, Map, Scientific Reports, Letters, etc.*, pp. 3, 9.

other men's coin. Based on a principle that animates many successful commercials today—those featuring popular personalities huckstering everything from detergents to pork sausage and aimed more at consumers trust in the celebrity than his product—respected public figures were sought for mining company directorates. Plainly, their addition to the board might lead investors to judge the company by the trustees it kept. And many mining promoters sallied forth to find judges, business leaders, politicians, and particularly Civil War generals (colonels, if it became necessary) who enjoyed the respect and honor of the multitudes affected by the war. Revealed in Colorado mining ventures, for instance, are the names of generals Adelbert Ames, Nathaniel B. Buford, John A. Logan, Benjamin Butler, John A. Dix, and Henry W. Slocum. In some cases these men took an active part in the organization and financing of a mining concern as a forthright business endeavor, but often a promoter would simply give them shares or an interest in the project entirely free of charge, or perhaps ask for a "conditional subscription," make them titular officers or trustees, and wager that their reputation would bolster the sales of stock. Since directors did not assume individual liability to stockholders, no obvious legal or financial snares detracted from the gift.[15]

Such figurehead directors added "caste" to an undertaking, even though the caustic editor of one mining journal saw them as "dummies of the highest order of 'style' and 'toneness' " and mere "stool pigeons to cheat the public." At the height of the Leadville excitement of 1879–1880, Cyrus McCormick learned (and it probably did not come as a revelation to him) that a group of promoters only "wanted the prestige of your name for a large amt. as an indorsement of their scheme which enabled them to at once bring in all the capital they needed." The custom of using decoy investors stirred the *American Mining Index* of 1866 to conjure up the "Hunkidora Silver Mining Company":

OFFICERS

President — Honorable Jeremiah Blowhard
Treasurer — Honorable Gideon Graball
Secretary — Honorable Simon Smoothface
General Superintendent — Miner Botch, Esquire

[15] George E. Vigouroux, ed., *Diary of a Mining Investor*, pp. 20, 50–55; Harry J. Newton, *Pitfalls of Mining Finance*, pp. 67–68.

DIRECTORS

Major General D. Bility
Brigadier General Melisher Cervis
Honorable Tim Trumpet, Member of Congress
Honorable Verily Hardup
Adjutant General P. Q. Lation

LEGAL ADVISORS

Sneak, Quibble, and Steele

Apparently there were enough of those "simpleminded souls" who invariably connected "honesty with respectability, worth with riches, and integrity with prominence" and who felt that first-class citizens would not affiliate with a doubtful venture, to encourage its usage among mining promoters. Even so, if a mining concern, decorated with men who had no responsible interest or duties in the enterprise, survived and profited, it frequently did so in spite of the figurehead officers and directors who had been used to publicize stocks and add a dash of propriety.[16]

Still, most promotional campaigns revolved around the nucleus of a prospectus. Fundamentally, this served as a preliminary statement of the mining enterprise, giving facts and figures on the proposal, and was meant to give basic information to investors, whether large capitalists or the buyers of a few shares of stock. Typically it contained a list of company officers and trustees, along with their business affiliations, if helpful; testimonials from Colorado residents, especially those who expected to profit from outside investment; mining engineers' reports on the richness of the property, or, at least on valuable mines near the property; and, not infrequently, a paean to the promise of Colorado or even the entire western mining region of the United States. While in most cases it seems that the statement was primarily the handiwork of the promoter, waxing eloquent over the child of his own creation, there were some individuals, especially in the large cities of the East, who specialized in writing prospectuses and other promotional literature. For instance, the Gold and Silver Producers' Company in New York claimed that it prepared and published pros-

---

[16] *Bullion* 6 (October 24, 1881): 416; Thomas Ewing to McCormick, December 8, 1879, Series 2A, Box 34, McCormick Papers; *American Mining Index* 3 (January 22, 1866): 8; Clifford, *Rocks in the Road*, pp. 256–57.

pectuses and procured reports, surveys, and legal opinions on mines—all at a "special rate"—as well as acting as a transfer agent between financiers and mines for sale. These pitchmen in print, said Henry B. Clifford, a mining expert and investor who sharply criticized promotional techniques, were "word-artists," few of whom had ever seen a mine, and they earned their commissions on a sliding scale of exaggeration: "the more fascinating the picture, the more praise and price for the artist. They boast their ability to promise large profits," he added, "without really saying anything upon which a grand jury could indict." [17]

Naturally, superlatives rolled easily off the tongues of promoters, and, intentionally or not, they painted a false or inaccurate picture of the mines. When Clifford recommended that an operator in New York be a little less exuberant and a little more precise in regard to his properties, he drew an indignant reply: "Why, man alive! I would not sell a share in a thousand years if I were to do as you suggest! I have to dress up this mining game as the people won't buy and I obtain my commission. I have tried 'plain fact' literature and was sued for office rent." Sometimes the "dress" he spoke of appeared to be a shabby disguise. The prospectus of the Silver Cliff Mining Company near Rosita puffed that the company could deliver silver bars of 950 fine at a cost not exceeding five dollars a ton, including all the expenses of mining, transportation, and reduction. On other occasions the investor was treated to a potpourri of embellishments and half-truths on nearly every aspect of the mining industry that might sell a mine. Witness this appeal of the King Solomon Mining Syndicate to investors around the turn of the century:

Show me a country that has no mines and I will show you a people sunk in the degradation of poverty, and poverty makes cowards of nations as well as it does of men. Mining has transformed more broken men and tramps into multi-millionaires and placed them in positions of great power and trust, than any other business. Eliminate the miner and you wipe out civilization. Without the mines you could not have a frying pan, or a spoon, or a hat pin.

---

[17] The Gold and Silver Producers' Company also published the New York mining journal *Bullion. Bullion* 1 (May 16, 1879): 4; Clifford, *Rocks in the Road*, p. 96.

Somewhat facetiously, the anonymous author of a novel on mining speculation had his windy promoter promise not once to describe Colorado as the "Centennial State, youngest daughter of the Union, Switzerland of America, call climate 'Italian,' skies "Blue as Venice's' or compare the air to 'champagne.' "[18]

For the sharp operator transforming any mining claim, including those unproved, inaccessible, or downright worthless, into seductive storehouses of instant wealth often became just an "intellectual feat." At times this meant no more than stressing the merits of the property while nimbly sidestepping its negative points, or, as indicated above, deftly associating the claim for sale with the romance and glory of centuries of mining. Beyond that it became a question of how much more the vendor could promise his buyers without actually slipping into deceptive practices or pure fantasy. Shortly after two men issued a bombastic brochure on a new mine at Cripple Creek about which they knew very little, pranksters in the clubs of Colorado Springs parodied their attempt in a prospectus for the Tenderfoot's Delight Mining, Milling and Transportation Company, which stated in part that,

The mine is elevated two miles above sea level and consequently the grade of ore expected to be found will be very high. A long flow of water exists in this great mine, which, if pumped to the surface to a canal which is projected to Colorado Springs, will form an important artery of commerce.

Though the object of ridicule in this case, the Cash on Delivery claim in Poverty Gulch, developed into a first-class producer in the gold camp of Cripple Creek, the lesson is not to be lost that dozens of other promotional statements confused and misled investors by their extravagant fictions.[19]

Reflecting on the widespread tendency of people to overestimate

[18] Clifford, *Rocks in the Road*, p. 109; Silver Cliff Mining Company, *Prospectus*, p. 11; King Solomon Mining Syndicate, *Opinions*, p. 4; "A Speculator," *A Speculation*, p. 5.

[19] The promoters of the C.O.D. in 1891 were Spencer Penrose and Charles L. Tutt, both of whom became wealthy from their mine and smelter investments (N. E. Guyot, "Cripple Creek: An Inside Story," *Engineering and Mining Journal-Press* 118 [December 13, 1924]: 937). John F. Graff, *"Graybeard's" Colorado: or, Notes on the Centennial State*, p. 71.

the values of gold or silver property, Clifford believed that the "very best of the mining population is given to exaggeration." What might be called unintentional deception in the sale of a mine frequently started with the prospector, who, in describing his claim to a purchaser or a promoter, related an overly optimistic and inaccurate picture of the property, obviously letting his own hopes and imagination run roughshod over the facts. Such examples led a New York newspaper to wonder piously about the Colorado climate, suggesting that the "air is so balmy and exhilarating that it is well-nigh impossible for the people to either talk or write as they do in the eastern states." Certainly some exuberant prospectors managed to see a four and one-half foot wide vein on their claim rather than the actual four feet or to add a few more feet to a lead of seventy-ounce ore that was really only fifty ounce. Instances also occurred when the discoverer thoughtlessly assayed high grade samples of his ore, then proudly—and mistakenly—announced that the mine had an unusually high "average value." In fact, a true estimate of value could be reached solely by careful testing, normally by a qualified mining engineer, of specimens taken at random from as much of the mine as possible. If this procedure had not been followed, false impressions about the claim arose and commonly reverberated throughout the sale and organization of a mining venture.[20]

In approximately the same category appeared the question of whether or not Colorado possessed mineral veins that became increasingly more valuable as miners followed them deeper into the earth's crust. This question, while a matter of some controversy among experts in mining geology, confused investors and made facile slogans for promoters. The issue focused on a conclusion reached early in the territory's history that Colorado had been favored with "true fissures," meaning lodes or veins of ore that grew in richness with depth. In 1863 a Nevadan who heard this stated as fact by nearly everyone he met in the Rockies, duly reported that "all the Pike's Peak folks have to do is to go down deep enough and all their mines will pay." Similarly, most prospectuses issued during the first twenty years of Colorado mining, and for many years thereafter, carried their lists of true-fissure veins and conveyed the impression that surface indications,

[20] Clifford, *Rocks in the Road*, pp. 45, 62–63; *New York Times*, June 19, 1873, p. 4; Balch, comp., *Mines . . . in 1882*, p. 855.

even outcrops of high-grade ore, would be small in comparison to the bonanzas at lower levels. Amasa McCoy, a professor of literature and music and a promoter for the International Mining and Exchange Company, lectured a group of investors in a Chicago opera house:

All of this company's mines I believe to be clear and certain cases of true and well defined fissure veins; and if you were to put three shifts of men to work upon them to-morrow, and ore were to be taken from them every day, and every night, for a century, so far from being then exhausted, your children and your children's children, down even to the third and fourth generations, would probably see these properties even more remunerative than you will ever see them yourselves.[21]

By their nature, true-fissure veins were supposedly neither superficial nor segmented; that is, they were not apt to vanish permanently into the surrounding rock or "peter out," in the parlance of the miner. Consequently eastern companies tended to spend great sums of money searching for a vein that had "temporarily" disappeared or, in the words of one mining engineer, foolishly "running after lost money." Basically to prevent such mistakes and to save the investors' dollar for more rewarding projects, some engineers, notably Thomas A. Rickard and Rossiter W. Raymond, generally denied the axiom of greater richness in depth, despite the stubborn resistance of the mine-promoting fraternity, whose purposes the theory served well. After Rickard, speaking as the state geologist of Colorado, told a University of Colorado audience in 1898 that lodes did not grow richer and that all men were not born free and equal, he recalled being damned by the promoters for puncturing their "glittering fallacy" and being denounced by the Grand Army of the Republic for being unpatriotic, yet he rejoiced in destroying "two delusions."[22]

21 The *Ouray Times* offered this definition of true fissures: "a term used in contradistinction to false fissures of minerals, which, after promising the sanguine holder a fortune equal to Jay Gould's, suddenly peters out and is no more, whereas your true fissure continues and if he persists will eventually crop out in China and Japan" (March 11, 1882). *MSP* 6 (April 20, 1863): 2; John Wetherbee, Jr., *A Brief Sketch of Colorado Territory and the Gold Mines of that Region*, p. 15; Amasa McCoy, *Mines and Mining in Colorado: A Conversational Lecture*, p. 30.

22 Another engineer, Winfield Scott Keyes, disagreed with his colleagues. At Ouray, he said, there were true fissures that "reach down as deep into the earth as the ingenuity of men will ever follow" *EMJ* 28 [August 16, 1879]: 113). Thomas A.

Gleeful though he was, Rickard and his fellow engineers must have realized that in terms of fashioning a public image of the Colorado mines and of the western industry, their lectures were often mere popguns compared to the heavy artillery of the promoter. For, in addition to the heady prospectuses that circulated widely and the personal exhortations to men of capital, some projectors tried to reach a mass market. They started with the public press—newspapers, popular magazines, and nonmining journals—by which might be tapped a larger, more varied, and less conservative investment market than manufacturers, merchants, and bankers, although they too would be included. If used with imagination, these periodicals could lend an air of authenticity and respectability to a mining venture that it might not otherwise enjoy from the fund-raising and tub-thumping character of a prospectus. Coordinating a clever campaign in 1896 to sell the stock of the Raven Gold Mining Company of Cripple Creek, J. Parker Whitney first "kept the Colorado papers somewhat warm" from his Boston office with "dozens" of stories he released about the mine; after these appeared in the western press, he then had them reprinted in eastern newspapers. In this roundabout way the articles became news items, eye-witnessed and unbiased, to be ingested by easterners along with columns on the McKinley-Bryan race for the presidency and the specter of free silver, while Whitney coyly waited in "readiness for a stock movement." Wherever skillful promoters operated in the East and West, they apparently wrote dispatches about their mines and submitted them to local newspapers, thus making their own news. Then, if everything worked smoothly, their properties received free advertising, took advantage of the newspaper's function as a reporter of fact, and still benefited from the promoter's choice of timing, detail, and coloring.[23]

In cases where the vendor wished to attract small or medium-sized investors, who did not usually merit personal courting by the promoter or receive mining brochures or regularly scan the pages of the business and financial weeklies, this type of mass publicity, tied

Rickard, *Retrospect: An Autobiography*, p. 68; Rossiter W. Raymond, *Statistics of Mines and Mining in the States and Territories West of the Rocky Mountains* (1870), p. 5; Etienne Ritter, *From Prospect to Mine*, p. 67.

[23] Clifford, *Rocks in the Road*, pp. 189–94; *Bullion* 5 (February 14, 1881): 51; Whitney to Eben Smith, October 10, 1896, Box 2, Smith Papers.

in with other methods, had special application. Frequently he aimed to sell so-called penny stocks, shares that ranged from a few cents to a dollar each in extremely speculative and hazardous mining enterprises. By advertising in the press and using a list of "poorer people" obtained from the credit agencies, a New York dealer successfully placed low-priced mining stock in the hands of clergymen, lawyers, physicians, teachers, and women. In the 1890s the Prudential Mining Investment Association of Florence (Colorado), proclaiming that "In Union There Is Strength," launched an investment pool for the "thousands of wage earners, clerks, mechanics, and others of moderate means" interested in the Cripple Creek mines, and produced a plan for "easy monthly installments" of five dollars down and five dollars a month to buy its fifty-dollar shares. To make contact with a similar clientele in the eastern cities, several promoters rented office space in a "semi-religious building," one in which religious books, articles, and church supplies were sold, and opened their trade among people of strong faith.[24]

Nor were rural areas neglected. Besides running effusive mining pieces in the local newspapers, some promoters hired special railroad cars in which they toured the countryside and peddled a few shares of stock at each way station or commissioned a town banker or merchant to "work" his neighbors. "Brick" Pomeroy may have used this method or one of its variations in 1881 when he was accused of a "grand steal" for selling ten-dollar shares in his Standard Mining Company to the "unsophisticated natives" of the Deep South, Oregon, and Nova Scotia. In the opinion of T. A. Rickard, the Middle West (particularly Iowa, Indiana, and Illinois), and other agricultural regions generally out of touch with the mines made "fertile fields for schemers," who placed mine advertisements in the farming and religious papers to catch what the *Engineering and Mining Journal* sneeringly called the "super-credulous class of persons who are prey for quack medicine vendors, prize puzzle concerns and $50 a week-at-home men." However, from such examples can be seen that side of mine promotion—often its most flagrantly fraudulent—which operated among the less informed classes of small investors and which, to an

[24] *MSP* 43 (September 24, 1881): 202; Prudential Mining Investment Association, *Prospectus*, pp. 3–5; Balch, comp., *Mines . . . in 1882*, p. 863.

extent, also recognized the financial potential of those groups whose slim savings, perhaps stashed away in rusting tin cans, were not usually worth the while of high financiers in manufacturing or transportation.[25]

Evidently the size and rewards of the market among smaller investors were ample enough to encourage some operators to expand and experiment further in that direction. Through cleverly contrived schemes aimed directly at the individual with a few dollars in savings and the person unfamiliar with the complex and exciting world of mine finance, promoters added new dimensions to their art. One firm, the United Mines Company, which was based in Denver and had a Chicago office on LaSalle Street, offered shares at three cents each to small investors who preferred a "promoting company, which should not risk any of its funds in mining, but should promote other companies to take these risks." From the interest United Mines would hold in these auxiliary or operating companies, spouted the promoters, "visions of wealth arise almost sufficient to turn the brain." Essentially, the company based its appeal on the popular notion that promoters, those "who arrange and conduct to a successful termination sales of mining properties," not purchasers, make the "great fortunes" in mining. Whether or not United Mines survived beyond its prospectus, its plan to become an investor-owned promotional agency and a mines-holding company was novel. The United States Mining Investment Company, formed in 1880, took a different tack. Headed by mining engineer Edward Bates Dorsey, the company boasted a gilt-edged directorate composed of eastern bankers and businessmen, including sugar refiner Henry Havemeyer, publisher Daniel S. Appleton, and lawyer Ulysses S. Grant, Jr.—"household words in our best financial circles," said one journal. The company intended to buy and sell mines, primarily around Leadville, once its engineering staff had examined and endorsed the properties for sale to the public. As with the United Mines, Dorsey's enterprise sounded the note that high returns on small investments were possible in mining, if cooperation and careful management could replace the usual hazards of mine

---

[25] *DAB*, s.v. "Pomeroy, Marcus"; *EMJ* 32 (November 12, 1881, and December 17, 1881): 326, 407; *EMJ* 65 (April 30, 1898): 526, Thomas A. Rickard, ed., *The Economics of Mining*, pp. 109–11.

finance and if the investor could receive sound advice before spending his money.[26]

On occasion the person seeking reliable information on the mines tumbled into the trap of the so-called mining bureau, which disguised its true promotional purpose behind a cover of impartial investment counseling. The promoter who took the trouble to paint a sign on his door and order some letterhead stationery could foist himself on the public as the director of a "Bureau for Mines and Mining Interests of North America," or the "Colorado Mining Bureau," or the "Bureau of Mining Information and Collections," the last being the brain child of a Silverton newspaperman, mine promoter, and bill collector. In 1877 the *Engineering and Mining Journal* reported that eastern cities such as New York, Chicago, and St. Louis harbored a number of fraudulent mining bureaus, operated by "two or three impecunious adventurers" and a "broken down chemist, which, when they could not pay their rent, melted down their ore samples and moved out of town." Of better reputation, though still pretentious and deceptive in its title, was the Colorado Mining Bureau, which functioned from an office on lower Broadway in New York and was managed by a promoter in association with David H. Moffat and Eben Smith, two of Colorado's leading mine owners and dealers. Under conditions such as these, it would seem certain that matters of fees and commissions as well as special pleading for given pieces of mining property, rather than the best interests and needs of the investor, became the chief concerns of the bureau.[27]

Privately owned mining-stock exchanges, which proliferated East and West to sell selected issues of securities, often stood accused of the same shady and misleading practices as the bureaus. Here again the small investor was the principal target of operators who sold low-priced stocks, ordinarily in companies of their own creation, and presented a legitimate front, when in fact all indications suggest that they were little more than gambling dens and breeding places for wildcat

[26] United Mines Company, *Prospectus*, pp. 2–4, 8–14; New York *Bullion*, comp., *Bullion: Its Production and Use*, back cover; *EMJ* 29 (January 10, 1880): 24–25.

[27] *EMJ* 18 (September 26, 1874): 196, 203; *EMJ* 24 (September 22, 1877); 219; L. V. Deforeest to David H. Moffat, January 30, 1897, Box 2, Smith Papers; C. S. York, ed., *San Juan's True Fissures*, p. 8.

schemes to abuse the "dear public." Before closing its doors in 1877, the Colorado Mining Stock Exchange had successfully snared many easterners into buying shares in a group of phantom mining companies conjured up by its director, one Moses Anker. Anker had achieved a measure of notoriety several years earlier for his sale of the Caribou silver mine, west of the town of Boulder, to a Dutch syndicate for three million dollars, estimated to be six to ten times its worth—which rated as "perhaps the greatest mining swindle on the continent." Fresh from his triumph in the Caribou transaction, Anker took over the exchange in Denver in 1876, a one-room office shared by a struggling tailor, and advised easterners through the newspapers to check with him before purchasing any Colorado mining stock, a course which many apparently followed to their pecuniary loss and embarrassment. "In nine cases out of ten," decried the *Engineering and Mining Journal* in regard to stock exchanges like Anker's, "their touch is contagious" and they need to be scrupulously watched by the investing public. Yet in the mid-1890s the same journal had reason to denounce the Skinner and Company stock exchange in Chicago for similar activities and accused the manager of peddling worthless Cripple Creek stock at two to fifteen cents a share, or a fraction above the cost of paper and ink used in printing the securities.[28]

The countless incidents of misdeeds by many promoters during the last fifty years of the nineteenth century gave substance to the popular image of the middleman as a sharper, a deceiver, even a thief, whose skulduggery injured thousands of honest investors. The devious means and methods used by the promoting crowd led one paper to conclude in 1880 that mining had become "almost a synonym for knavery and uncertainty." Mark Twain contributed his sarcastic definition of a mine as a "hole in the ground owned by a liar," to which a mining engineer added, a "hole in the ground sold by a lying promoter to a stupid investor." Much of this sorry reputation the promoter earned, and it received nearly weekly confirmation in the newspapers, mining journals, and other records of the era, which carried tales of intricate webs spun to catch the "tenderfoot" investor. There was, for instance, the Rocky Mountain Prospecting and Mining Company—organized by two prospectors, one of whom professed to be a

---

[28] Newton, *Pitfalls of Mining Finance*, pp. 114–15; *EMJ*, 23 (April 21, 1877): 252; *EMJ* 24 (August 11, 1877): 105–07; *EMJ* 57 (May 19, 1894): 458.

deputy United States surveyor and a mineralogist—which claimed to own twenty-two gold and silver mines that could be purchased for as little as one hundred dollars each. Luckily, the local Colorado newspaper, the *Rosita Index*, denounced the prospectors as the plan appeared in New York, exposed the leaders as notorious swindlers, and denied that the pair had twenty-two mines or even two hundred and fifty cents between them.[29]

Advance warning might have saved more of the money invested by easterners in the mining proposals of a self-styled "professor," James Cherrie, who, in the words of a Denver newspaper, followed a career of "deceit, robbery and dissipation." Cherrie, reputedly a well-dressed and fast-talking promoter, versed in all of the "geological and meterological" terms of the day, steered several Rock Island Railroad executives into an Arizona mining venture that cost them between $10,000 and $20,000. Never able to find the mine they supposedly owned, the railroad men forced Cherrie to repay the money, but, undeterred, the "professor" next interested a group of Chicago capitalists in his Eclipse (correctly named!) Mining and Milling Company, situated in the San Juan region of Colorado. As manager of the enterprise, he expended over $50,000, kept no records, and ran the company into debt without showing any significant development work and, of course, without returning a dividend. At the time of Cherrie's exposure in the Chicago and Denver newspapers of 1882, New Yorkers were also accusing him of swindling some $250,000 from them in the sale of two Colorado mines. It seemed likely that the irrepressible Cherrie, once he was released from a New York jail, would simply find another city in which he was unknown to repeat his scheming.[30]

For sheer boldness, few examples of fraud in the early Colorado mining industry surpassed the "dying miner" routine used to delude

[29] Apparently wagering on public sympathy for the reformed swindler, one Bostonian promoted his company by admitting that he had previously "manipulated" a mine and had once been an "exploiterer," but, having changed his evil ways, he now deserved full "confidence and consideration" (John Wetherbee, Jr., *A Letter on Colorado Matters to the Stockholders of Excelsior Co. and others whom it may concern*, p. 17). *New York Times*, June 19, 1880, p. 4; *EMJ* 25 (February 16, 1878): 114; John Hays Hammond, "Suggestions Regarding Mining Investments," *EMJ* 89 (January 1, 1910): 11.

[30] *Denver Tribune*, April 8, 1882.

several men of capital in New York in the 1890s. Principal losers in the hoax were George C. Upshur and his plant manager, a Mr. Rogers. An associate of Comstock millionaire James C. Flood and a highly successful steel manufacturer in his own right, Upshur was no stranger to the manipulations of mining. The elaborate ruse began innocently when Rogers, returning to New York on the night train from Pittsburgh, befriended an old miner, Harry Belmont, who, from all indications, neared death in an adjoining berth. Solicitously, Rogers helped the fellow to a New York hotel, where he learned that the prospector owned a mine in Clear Creek County; worth an estimated $3 million, the mine was currently threatened by a law suit from a neighboring claim holder. Implored by Belmont to help him save the property for a younger sister in the Middle West, the New York businessman agreed to do what he could in arranging a settlement of the dispute in return for a share in the mine. With $20,000 from the old man, Rogers set out to purchase the adjoining claim from a prospector named Henry E. Miller, who, perchance, had moved into another hotel in the city. However, Miller drove a hard bargain, insisting upon $60,000 for his property, and the steelman turned to Upshur for help in raising the additional money, no doubt pointing out to his employer the ease with which they could acquire control over an immensely rich mine. Upshur responded quickly, contacted Samuel M. Roosevelt, portraitist and cousin of Theodore, and the three men managed to locate the funds; subsequently, Rogers and Upshur accompanied Belmont, still manfully warding off death, and Miller to Denver to close the transaction and visit their new silver property. All seemed satisfied: Miller had his $60,000, the New York capitalists shared a mine, and Belmont was at last ready to die in peace. Overnight, Belmont secretly checked out of his boardinghouse, leaving behind a trunk full of bricks but no deed or location for his wonderful mine. The grand swindle had been completed.[31]

At first Upshur and his embarrassed partners agreed to keep silent about the hoax so that friends would not enjoy the "screaming joke" played on them, but six months later the owner of New York's luxurious Grenoble Hotel was gulled by the same people. William Noble, the proprietor, had collected $100,000 and taken ten friends

[31] George L. Upshur, *As I Recall Them: Memories of Crowded Years*, pp. 204–07.

to Colorado with him in a private railroad car to inspect the mine in Clear Creek County. Although this time Belmont's "dead body" was actually hauled from his room, he again left behind the pile of old red bricks and nothing more. Only after hearing this did Upshur disclose his own unfortunate experience and allow "our friends to have a laugh at our expense."[32]

In the day-to-day business of buying and selling mines and mining stocks, the capitalist had always to be on his guard against subtler, less theatrical performances than those related by Upshur. Actually the dishonest promoter depended upon a basic repertoire of tricks, limited more by his capacity to deceive and by the tried and true methods of the past than by any legal barriers erected in this era of freewheeling finance. For instance, an operator in possession of a poor or undeveloped claim might give it the same or nearly the same name as a great and prosperous western mine, if in doing so he might throw dust (and probably not gold dust at that) into the eyes of distant investors and loosen their purse strings. Taking into account prospectors who innocently flattered their holes in the ground with famous names, it happened too often to be purposeless. The practice, said one journal, bordered on the "criminal." A Colorado mining directory for 1879 listed seven Comstock, six Homestake, five Potosi, and two Gould and Curry mines, each a rich property outside the state, but likely to mislead and confound the easterner who knew the name and the reputation but nothing more.[33]

Even the most discreet capitalist seeking some tangible evidence of a mine's worth before he invested might be hoodwinked by the specimen fraud. Acting either as the chief promoter of a mine himself or renting out his treasure to another vendor, this character specialized in contriving an elaborate and persuasive display of choice ore samples, meticulously arranged to show the streaks of gold or silver and nestled deep on a tray of velvet. For maximum effect, the specimen might be surrounded by bars of purest gold and silver, readily fashioned out of melted coins. After dazzling prospective customers

[32] Ibid., pp. 207–08.
[33] Thomas B. Corbett, *The Colorado Directory of Mines*, s.v. "Comstock," "Gould and Curry," "Homestake," "Potosi"; *MSP* 43 (October 22, 1881): 265; *Mining Record* 6 (October 11, 1879): 291; *Financial and Mining Record* 31 (January 9, 1892): 22.

in one venture, the exhibit could then be returned to its owner, the specimens chipped and filed into new configurations, and the ingots recast for duty in another sale, with another promoter and perhaps in a different city. In the early 1880s it was alleged that both a Colorado and a New Mexico mining company, neighbors in a New York office building, conveniently used the same ore samples—taken from a mine in Utah.[34]

To conclude from these examples that all promotion affecting the Rocky Mountain mines or that mining promoters as a group were somehow uniquely dishonest or even depraved would be unfair and inaccurate. In fact, the entire western mining industry, from the copper deposits of Montana to the silver lodes of Nevada, bred treachery and double dealing throughout most of its history, and promoters in all fields after the Civil War ranked low in terms of ethical conduct, as did many of their social betters such as the Conklings in politics and the Goulds and Vanderbilts in big business. Without excusing the gross exaggerations, the half-truths, and the outright frauds attributed to mining promotion, it should also be made clear that the apparent overstatements that became truths were often ignored or adopted as testimony to the vast richness of the West, while the hyperbole that proved false was censured in the press and elsewhere. A newspaperman in Leadville said it differently: "everything here that succeeds is considered legitimate." Furthermore, when an investment in some Colorado mine failed due to mismanagement, weak financing, greed, or other reasons, it was a natural tendency for capitalists to lay the blame on that individual who took them into the venture in the first place. Thus, the promoter became a convenient scapegoat for deeper problems within the industry or for the blunders of absentee mine owners.[35]

It was an article of faith among large numbers of Americans in the last half of the nineteenth century that mining for precious metals was an unobstructed road to wealth. From a small investment could grow a huge fortune almost overnight. Experienced mining men knew better, however, and realized that there were tremendous risks in the business. They held to the adage that nobody could see into a mine

[34] Frederick H. Smith, *Rocks, Minerals and Stocks*, pp. 222–23; Balch, comp., *Mines . . . in 1882*, p. 863.

[35] Graff, *"Graybeard's" Colorado*, p. 71.

beyond the last drill hole, meaning that either a great bonanza or merely tons of waste rock might lie ahead and that the current state of mining science left much to be desired in reducing this level of uncertainty. Between the ignorance and greed of investors and the real vagaries of mining stood the promoter, and he served his interests best by selling a piece of property, not in acknowledging or computing the odds against successful investment. Under these circumstances, misleading statements from promoters gained momentum and accelerated sharply in boom times when more anxious customers (with more capital, and knowing less about mining) and more disreputable vendors (with more properties, and caring less about their methods) came together, resulting in fraud and lost money. All of which led a mining paper to philosophize that "fungi will always be found upon the richest soil, and parasites will ever thrive upon the most nutritious vegetation."[36]

The great distance separating property from potential buyer added complexities of a different sort. Though a promoter might effect the sale of a mine hundreds of miles away sight unseen, he might also have felt the need to pressure an investor more, normally by making high-blown claims for his property, since slow negotiations over a distant mine might cause capitalists to look closer to home for an attractive investment. At the same time, far distant properties, reasoned the *New York Times*, left the buyer at the mercy of a promoter whose honesty and good judgment alone stood between success and failure, and the newspaper deplored the "appeal to credulity" that necessarily followed: "The transaction is one where individual judgment is eliminated; it is an exercise of truth in others; it is a ticket in a lottery where there may be no prize and not even the intention to have a prize." Furthermore, if it were completely true that in mining, as in art, distance lends enchantment to the view, then promoters might have painted fewer marvelous word-pictures of the mines they hoped to peddle. Actually promoters sometimes evoked the grandest images of the romance and abundance of mining to hurry along a transaction stalled by a financier's concern over a remote investment. This seemed to be so in 1867, when Cyrus McCormick, not one to be rushed, questioned the strict accuracy of a few state-

[36] Newton, *Pitfalls of Mining Finance*, p. 23; *MSP* 40 (February 14, 1880): 104.

ments in the prospectus of the "Equitable Mining Company of Chicago in Colorado." It was not "very material," answered the promoter
impatiently, "whether or not all there stated be strictly true," for it
was "mainly true," he announced, that *there is* [sic] *rich gold and
silver mines in Colorado.*"[37]

Livelier competition between promoters to place their wares in
the East usually attended the opening of rich new mining fields in
Colorado or elsewhere in the West. In the rush to outtalk a rival,
misinformation and exaggeration flowed freely, and the buyer was
likely to hear most anything but an accurate and unvarnished description of a mine. After all, any good salesman knew that in a period of
high promotional activity the success of his own presentation to
financiers depended on his being so utterly convincing that they would
be hooked and reeled in before the next bait floated by. In the heat
of excitement over the silver discoveries at Leadville, vendors in the
East reportedly made "ill-natured remarks" about each other and attempted in every way possible to discredit their fellows and sabotage
their efforts, even as other rumors told of promoters being secretly
paid by their competitors to keep silent and not obstruct pending
sales. Until a mine could be delivered into the hands of its new
owners, the promoter usually worked in fear that some individual or
circumstance would arise to confound the transaction. In the 1890s,
Henry R. Wolcott, a prominent Colorado businessman, may have "un-
intentionally" discouraged the purchase of a mine by Boston banker
Henry Lee Higginson. According to the promoter in the operation,
Wolcott, himself having come East to trade a mine or two, disparaged
the other property in a conversation with Higginson. Observing that
this talk may well have doomed any deal with the banker, the vendor
complained that people like Wolcott will often destroy confidence in
a property without knowing it is for sale, though he supposed that
this was the occupational hazard of the promoter. In view of what
had happened, he concluded—as other promoters must have—that in
selling frontier mines it "always takes a pound of argument to allay
an ounce of suspicion."[38]

[37] *New York Times,* June 19, 1880, p. 4; James Lamson to McCormick, January 10, 1867, Series 2A, Box 34, McCormick Papers.

[38] *MSP* 39 (July 26, 1879, and December 27, 1879): 56, 410; W. H. Brevoort to Eben Smith, April 19, 1897, Box 2, Smith Papers.

For better or worse, the mining promoter played an indispensable role in awakening eastern capital to the natural wealth of Colorado and the region's economic potential. Purely in terms of his accomplishments as a financial intermediary, the dealer in mines brought money and interest to a new and struggling industry that could not have grown or survived without large injections of absentee capital. To one student of this process, the promoter was no less than a "John the Baptist of industrial development." Ultimately, of course, they promoted not solely the mines but the state and the Rocky Mountain region as a whole by focusing attention on the real opportunities and rewards, and they accelerated the economic advancement of the West by channeling funds into remote and undeveloped areas. Grudgingly, one tough critic of mining promoters conceded that even the most dishonest among them "managed to bring enough gold to Colorado and their own pockets to make the country the El Dorado they had pictured."[39]

Whether amateurs or seasoned professionals, easterners or Coloradans, most mining promoters would have recognized themselves in the parody of "Smiles" sung at a meeting of mining men in 1920:

> There are mines that make us happy,
> There are mines that make us blue,
> There are mines that steal away the tear drops
> As the sunbeams steal away the dew.
> There are mines that have the ore chutes faulted,
> Where the ore's forever lost to view,
> But the mines that fill my heart with sunshine,
> Are the mines that I sold to you.

How many eastern mine investors may have been amused is not known.[40]

[39] *MSP* 32 (February 19, 1876): 120; Newton, *Pitfalls of Mining Finance,* p. 153; Samuel S. Wallihan and T. O. Bigney, *The Rocky Mountain Directory and Colorado Gazetteer for 1871,* p. 151.
[40] *Mining and Metallurgy* 1 (April, 1920): 6.

3

Financiers of the Frontier

I<small>T</small> was the financier, not the romantic old prospector or the quick-talking promoter, who ultimately built a hardrock industry in the rugged and remote mountains of Colorado. What the prospector discovered high in the Rockies and the promoter tried to peddle in towns and cities across the country would have certainly remained an undeveloped mineral resource without the millions of dollars invested by easterners, middle westerners, and other distant and diverse capitalists, large and small alike. In capital lay the means of properly opening a mine, sinking the shafts, timbering it against sudden shifts of dirt and rock, and draining it dry. And capital made possible the hauling, milling, and treating of tons of ore, as well as a multitude of steps and processes required to produce bullion from the raw material. As many a poor miner came to realize, it took money to make money on the Colorado mining frontier. Stressing this fact in 1898, T. A. Rickard informed the readership of an important financial journal that success in mining, like Napoleon's formula in war, depended upon three fundamentals: "the first is money, the second is money, and the third is money." And repeating the old Mexican proverb that a gold mine is needed to work a silver mine, the engineer quipped that it usually "requires the aid of a bank to develop a gold mine."[1]

If Rickard seemed to be overstating the difficulties and costliness of deep-level mining in Colorado, he did so with the best of intentions and after years of experience in the region. Since the dawning of the industry in the 1860s, a sizable portion of the investing public had held mistaken and sometimes damaging ideas about the business

---

[1] Thomas A. Rickard in "Colorado: Resources and Attractions of the State," *Bankers' Magazine* 57 (July, 1898): 191.

of mining. On one hand there was the persistent belief that mining involved nothing more than digging gold and silver out of the earth in much the same manner as a farmer harvested his potato patch, albeit with one significant difference—a crop of precious metal made the prospects infinitely more exciting than a sack full of tubers. On the other hand, mining was reduced in the popular mind to a game of pure chance, a gambling session, an elementary form of speculation in which the ordinary rules of sound business conduct did not apply and in which a fortune could be made on the stock exchanges. Or, with less effort than "playing" the stock market, the mine itself might actually prove a bonanza and pour immoderate sums of money into the bank accounts of its lucky owners. Both schools of thought had their adherents, and in either case the western mines appealed to the basic urge to get-rich-quick and to get-rich-easy. No wonder promoters joked that the richest mine could always be "found in the purse of a fool."[2]

Dumped in the same category with games of chance and spinning roulette wheels, mining operations became to some critics vulgar debauchers of the good morals and character of the American people. At times mining was both condemned as "inimical to the welfare of the people" owing to its fluctuating and hazardous nature and charged with "contempt" for the "more gradual methods of acquiring wealth." Journalist J. F. Graff solemnly warned his readers against seeking a fortune in the mines, where the chance of becoming a "moral shipwreck, or of making your gain out of somebody else's loss" appeared much greater than in a good, solid investment in Iowa farm land. He also hinted that the wealth gained from a lucky strike in the Colorado mines favored the investor and the nation less than a bountiful crop of corn raised on the fertile fields of the Middle West. How much impact disapproving words of this kind had on the conduct of the investing community is hard to gauge. Most likely investors shaped their reaction to the western mines around sterner stuff, around personal and general financial considerations, and, where concessions were made to moral scruples, an investment may have been

[2] The *MSP* thought there were large numbers of people who believed that a "mine should pay as soon as a pick is stuck into it, which would be like a building paying rent when the foundation is laid" (45 [November 18, 1882]: 329). John M. Stuart, *Mining: Its Theory and Practice*, p. 8; *EMJ* 61 (March 7, 1896): 214.

secret or at least unobtrusive. For instance, a clergyman's wife in New York concealed her venture because a minister's family is "subject to so much gossip that the least people know the better." On balance, the spirit of adventure and the promise of great wealth attracted a broad spectrum of American political, business, and military leaders as well as thousands of people in all social and economic groups to mine finance for varying lengths of time, to greater and lesser extents, and for high and low purposes.[3]

Throughout the late nineteenth century the names of prominent American political figures graced the stockholder lists of Colorado mining companies, which had an uneven record of success. In 1880 and 1881 Charles Francis Adams, Jr., bought some stock in the Chrysolite mine at Leadville and another "silver mine gamble" in Summit County, both of which failed soon thereafter and left the Boston Brahmin poorer by nearly $10,000. Bitterly and somewhat impetuously he confided to his diary that "I cheerfully accept the loss as the price of my emancipation from all faith in mining stocks, or information concerning them or any man who deals in them; and also the close of my operations in stocks." Another New Englander, James G. Blaine, the popular Plumed Knight of the Republicans and that party's presidential nominee in 1884, dealt rather extensively in Colorado mining properties for approximately twenty years, until his death in 1893. One of his early ventures began in the summer of 1879 when he joined with the former governor of Massachusetts and mayor of Boston, Alexander H. Rice, and an excongressman to purchase the Dunkin silver mine at Leadville for $300,000, which they then organized into a $5 million company. Between 1881 and 1884, the Dunkin reportedly paid dividends of $130,000, and it was eventually sold at a profit by Blaine and his partners to an eastern syndicate.[4]

---

[3] John F. Graff, "*Graybeard's*" *Colorado: or, Notes on the Centennial State*, p. 19; Mary Dickinson to Prince, September 22, 1894, LeBaron Bradford Prince Papers, New Mexico State Records and Archives Center, Santa Fe; T. H. Lowe, *Colorado: Its Mineral and Agricultural Resources*, p. 11.

[4] Diary entries for August 13, September 15, 1880, and June 24, December 9, 15, 1881, Charles Francis Adams, Jr., Papers, Massachusetts Historical Society, Boston; Frank Fossett, *Colorado, Its Gold and Silver Mines, Farms and Stock Ranges, and Health and Pleasure Resorts*, pp. 456, 575; Joseph G. Martin, *A Century of Finance: History of the Boston Stock and Money Markets, January, 1798, to January, 1898*, p. 230.

Over the same time period, a host of successful businessmen, many of whom were generally regarded as conservative in financial affairs—including credit authority Robert G. Dun, meatpacker John Cudahy, merchant Marshall Field, brewer Adolphus Busch, New York broker Jules S. Bache, and others who had earned national stature and had accumulated years of experience in business dealings —vigorously bought and sold shares in the mines. Often a mining venture or two in Colorado gave new depth or dimension to a larger portfolio of stocks and interests in western projects. William H. Russell fit this description. Beyond his partnership in the frontier freighting firm of Russell, Majors, and Waddell and his creation of the legendary Pony Express, Russell acquired the title of "father of Clear Creek County" because of his mining operations in the area. Having organized the Pike's Peak Express Company in 1859, Russell hauled tons of machinery and supplies into the mountains west of Denver and in the process gained a first-hand knowledge of the region's resources, which he then used in selecting the better mining claims for purchase. In 1863–1864 he spent some $50,000 in developing mineral properties around Idaho Springs, formed and promoted several companies at the neighboring camp of Empire, and with equal dispatch set up a score of operations in Gilpin County to work the gold deposits there. At a slightly less hectic pace, John A. Creighton, who had helped to construct the Western Union Telegraph system, used Omaha as his base to manage investments in western banks, real estate, railroads, stock raising, and the Colorado mines, which together accounted for his fortune of several million dollars. Although Russell and Creighton lived and traveled extensively in the Rockies, investors such as Levi Z. Leiter directed most of their frontier business from an eastern office. The merchant partner of Marshall Field in Chicago until 1881, Leiter invested heavily in the range-cattle industry and was the principal owner of the Iron silver mine at Leadville, recognized as one of the richest properties in the Carbonate camp.[5]

It was hardly surprising, in view of their ephemeral reputation,

[5] *EMJ* 25 (April 13, 1878): 257; *EMJ* 29 (June 5, 1880): 396; *EMJ* 31 (May 28, 1881): 372; *EMJ* 49 (February 15, 1890): 206; *EMJ* 64 (July 10, 1897: 44; Frank Fossett, *Colorado: A Historical, Descriptive and Statistical Work on the Rocky Mountain Gold and Silver Mining Regions*, pp. 87–93; Leiter Account Books, 1880–1886, Levi Leiter Papers, Chicago Historical Society.

that the mines attracted a colorful class of adventurers who added more tinsel to the financial end of the business through their antics. One of these was "Colonel" Jack Haverly, leader of the Mastodon Minstrels, which he advertised with cries of "Forty, Count 'Em! Forty!" and packed crowds into a dozen or more theaters he owned from New York to San Francisco. In 1880 he reportedly "bought up just about everything" around Gunnison and Irwin in southwestern Colorado, which included several shallow holes in the ground at a cost of between $100,000 and $200,000, ranches, sawmills, some real estate, and a newspaper, leading local people to wonder what would "appease such an appetite." Apparently more adept at promoting than developing his mineral properties, Haverly tried to sell them from his private stock exchange in Chicago and as he and his blackface troup toured the United States and Europe. An inveterate gambler, he reputedly made and lost five fortunes during his lifetime in poker games and stock speculations—at one time trying to seize control of the New York Stock Exchange—and eventually died of typhoid. At the time of his death, he was an indigent and unknown prospector near Salt Lake City. Similarly T. A. Rickard recalled that he had once managed a mine in Colorado for another sharp character, H. H. Warner, concocter of a vile tonic called Warner's Safe Cure, who merely "regarded mines as counters in a game of promotion and an excuse for a flamboyant finance that fooled the unwary."[6]

With high risk necessarily a part of the frontier mining industry, all investors, whether hot-blooded adventurers or sober members of the business establishment, became gamblers and flirted with an instant fortune against very great odds. One company prospectus almost cheerfully acknowledged that the speculative element was supreme, admitting that mining was "essentially a lottery, with a few great prizes and very many blanks. Every intelligent man who puts a dollar into a mine knows that it is a 'gamble,' but the spirit of adventure is strong, and there are always plenty of people to take the chances." J. Parker Whitney, who operated among eastern investors for forty years, observed that the mines fascinated men as nothing else did and often resulted in a "credulity and easy confidence not allowed in the

    [6] DAB, s.v. "Haverly, Christopher"; Frank Hall, History of the State of Colorado, 4: 150; EMJ 29 (June 19, 1880): 421; Robert E. Strahorn, Gunnison and San Juan, pp. 10–11; Thomas A. Rickard, Retrospect: An Autobiography, p. 53.

consideration of other business subjects." Time after time, conservative and experienced businessmen, who might have rejected highly speculative ventures in other directions, found the gamble and glamor of the mines irresistible and at least temporarily strayed away from their everyday pursuits and plunged into mining. Even as Norvin E. Green, president of the Western Union Telegraph Company, pledged it his "duty" to warn unsuspecting friends and relatives away from mining finance and grumbled that the "mines in Colorado were as uncertain as lottery tickets and the chances ninety-nine in a hundred against an investment in them," he confessed to having "sunk" over $100,000 in more than twenty mineral properties in the state.[7]

The great distance that separated an eastern investor from his western property accounted for much of the riskiness in mining. Communications between the two regions being slow and unreliable, the task of keeping abreast of daily developments at the mine and acquiring information from managers on the site always proved difficult. Without question, the problem of transmitting instructions and carrying on other business affairs over hundreds of miles hampered the flow of capital west, yet, measured against what they thought would be the probable rewards of a bonanza in gold or silver, easterners continued to invest, though sometimes wholly in the dark regarding their property. No less commanding a figure in New York business circles than the president of the Equitable Life Assurance Society, Henry B. Hyde, had doubts about the specific location and even the existence of a Colorado mine in which he held six thousand shares, and in obvious distress he confided to a fellow stockholder that "next time we buy a silver mine some of us better go out there and see where it is." Appropriately, the stock causing Hyde's discomfort bore the name of the Maid of the Mist Silver Mining Company.[8]

Nearly as numerous as the pitfalls of mining finance was the variety of avenues open to the investor who wished to buy a share of the Colorado mines. Particularly if he were a man of means, there were, besides the stock exchanges that sprang up in the major cities,

7 J. Parker Whitney, *Reminiscences of a Sportsman*, p. 400; Green to "Dear Susie," July 15, 1882, Letterbook 2, Norvin E. Green Papers, The Filson Club, Louisville, Ky.; *The Clear Creek Placer Company, Colorado*, p. 1, in Prince Papers.

8 Hyde to General Louis A. Fitzgerald, March 11, 1880, and February 12, 1883, Henry B. Hyde Papers, Baker Library, Harvard University, Cambridge, Mass.

likely to be proposals made by friends, relatives, and business associates, as well as unknown promoters seeking his help in financing
a new mining scheme, which they in turn may have joined at the
urging of others. This haphazard, informal process accounted for the
largest share of Colorado mining investments made in the days before
a systematic investment market developed. Consequently, in circuitous
and complex ways, a capitalist often became tied to a hole in the
ground miles away in the Rock Mountains about which he knew little
and suspected less. For example, the course followed by one of the
most famous and successful investments in Colorado, that of merchant
Meyer Guggenheim, began with personal contacts and loaned money
and resulted in a bonanza of silver and the start of an enormous family fortune. In 1879 the discoverer of the A.Y. and Minnie mine on
Iron Hill at Leadville sold out to a pair of Philadelphia natives, Samuel
Harsh and George Work, who soon accepted as partners two more
Philadelphians, Thomas Wier and Charles H. Graham, all of whom
had come to Colorado searching for an opportunity in the mines.
With the exception of Harsh, the group had signed promissory notes
to purchase the property and as the due date neared on the paper
Graham hurried off to Philadelphia in hopes of securing a loan or
selling a piece of the mine to raise the money. After contacting several persons in the East, he approached an old friend and fellow
merchant from the Civil War years, Meyer Guggenheim, already
quite wealthy, who was apparently intrigued by the scheme and extended a loan to Graham in return for an option to buy into the mine.
As development of the mine advanced—made possible by the new
funds—revealing a deposit of high-quality silver ore, the former ped-
dler of stove polish bought the shares of Work and Wier for $5,000,
and by the mid-1880s Guggenheim had acquired a controlling interest
in the property and was reaping the rewards of a very fortuitous investment. For a while his mine returned $2,000 in silver each day and
yielded over nine million ounces of the white metal. It also set into
motion a remarkable family which in the twentieth century exercised
unparalled control over the western mineral industries through the
American Smelting and Refining Company.[9]

A far more detailed look at how eastern money found its way in-

[9] Donald F. Popham, "The Early Activities of the Guggenheims in Colorado,"
Colorado Magazine 27 (October, 1950): 264–65; Harvey O'Connor, *The Guggen-*

to Colorado appears in the memoir of David Marks Hyman, a Cincinnati attorney at the time of his first venture. Born in a small
Bavarian village in 1846, Hyman had been brought to the United
States by his uncles, prosperous merchants in Cincinnati, who provided for the youth's early education and saw his graduation from
Harvard Law School in 1869. Along with other Middle Western cities
in the postwar decade, Cincinnati flourished in the 1870s, and the
fast pace of its commercial expansion enabled this personable and
capable young lawyer to develop a lucrative practice dealing principally in questions of business law. His fees were good—at one point
he received $150,000 for a case he argued—and Hyman was able to
put aside some money for the future needs of his growing family and
to make several investments, mostly in real estate and buildings
around the city. By 1879 he estimated that he had smartly invested
$40,000 and planned to expand upon these "paying" propositions as
long as his law practice met with continued success.[10]

But 1879 also brought exciting news from the West that large
deposits of silver ore had been discovered atop the Continental Divide
at a place oddly named Leadville, and Cincinnati soon hummed with
stories about easterners who had gone to the camp and quickly
hit pay dirt by poking a few holes in the carbonate beds or by purchasing a first-rate claim from the throng of penniless prospectors
known to inhabit the region. Typically blown out of proportion as
they filtered east, these tales nonetheless intrigued Cincinnatians as
they did people everywhere, and, along with many others, one Charles
A. Hallam decided to close his affairs in Ohio and head for Colorado
to "seek his fortune." While completing arrangements for his trip,
Hallam, who had been friendly with Hyman since they cooperated
in a law suit, asked the attorney for financial support in developing a
mine, should he stumble across a rich one in the Rockies. Though
dubious and a bit reluctant at first, Hyman had faith in Hallam's
"ability and integrity" and assented to his friend's request that $5,000
be made available on draft if and when opportunity knocked. Thirty-
six years later Hyman reminisced that he "had no idea of engaging in

heims: Making of an American Dynasty, pp. 27–28, 51–55, 60–62; Thomas A.
Rickard, *A History of American Mining*, p. 125.
    [10] David M. Hyman, "The Romance of a Mining Venture" (1916), pp. 1–10
(typescript memoir lent to the author by the late Donald M. Hyman, New York).

mining but thought I could take a little flier in the way of speculation and either make or lose in the neighborhood of $5,000."[11]

Once in Denver, Hallam supposedly studied the various mining regions of the state for some time in order to determine the best location for starting his new career and investing the money of his associate. It is likely that this research included conversations with experienced mining men around Denver and probably a few recently returned from Leadville, as well as with the ever present promoters who sensed he was in the market for a mineral property. There was, after all, no surer magnet for mining proposals than the scent of eastern money, and this probably brought Hallam into touch with "Professor" B. Clark Wheeler, a sharp-witted promoter who had come to Colorado a few years earlier. Educated at the University of Pennsylvania, Wheeler taught school for a while, was principal at several academies, earned a law degree, and, during a journey from British Columbia to Mexico in the 1860s, mysteriously acquired the professions of civil engineering and mining geology. Of paramount interest to Hallam, however, was Wheeler's knowledge of an area over the range from Leadville that he vowed would outshine and outproduce the booming camp along California Gulch. In Wheeler's opinion the recently opened Roaring Fork District of Gunnison County afforded the best opportunity to invest Hyman's capital before the price of claims rose beyond their reach.[12]

Thus at the close of 1879, Wheeler and Hallam interviewed groups of prospectors from the area around present-day Aspen who convinced them that silver lodes of tremendous value had indeed been discovered there. A few weeks later the pair took a bond and lease on the Smuggler, Durant, One Thousand and One, and several other claims on Aspen Mountain, handing over $5,000 in cash and agreeing to pay $160,000 more to complete the transaction by June 1, 1880. Hyman knew nothing of the purchase until he received a cryptic telegram from his agents on January 22 announcing that, "There is a flood tide in the affairs of men which if taken at its flood, leads on to fortune. We congratulate you. We have drawn upon you at sight for five thousand dollars." The attorney was "horrified" by the news and became increasingly distressed as he learned more of the details, par-

[11] Ibid., pp. 11–12.
[12] Frank L. Wentworth, *Aspen on the Roaring Fork*, pp. 30–31.

ticularly on the remoteness of the properties, the absence of railroad connections, and the heavy snows that visited the region. Still, content to "trust the luck," he undertook the chore of raising $160,000.[13]

Unexpectedly, and before Hyman had a chance to test the idea of co-ownership in a mine on his friends and business associates, he received an offer of help from an agent of Abel D. Breed, a Cincinnati casket manufacturer, patent-medicine man, and owner of the Caribou mine near Boulder when it was sold to a Dutch syndicate at an exorbitant price. Perhaps out of the million dollars he had made in that transaction, Breed gave Hyman a draft for $16,000 and a promise of additional capital and an advertising campaign to bring in small investors in return for a one-third share in the mining property. Breed's lawyer, George Hoadly, a former judge and soon to be governor of Ohio, filled Cincinnati with stories of the magnificent wealth of Aspen, told with the flourish and zeal of a politician, and began to greet his colleague on the bar as "David Millionaire Hyman," all as a part of the promotion. Regardless of the uneasiness much of this ballyhoo caused, Hyman conceded the advantages: he now believed he could meet the purchase price of the mine without further delay and encountered "no difficulty whatsoever in enlisting numerous friends who were all eager to share in my good fortune."[14]

Hyman's optimism was short-lived. Although by February, 1880, he had firmly committed himself to the investment at Aspen, ahead lay more than a year of wrestling with financial problems. When a report submitted by engineer John B. Farish questioned the value and the legal title to some of the claims, Breed angrily cut short his publicity drive for the property and sold half of his interest in it back to Hyman for $8,000. A further note of discouragement came from Hallam, who had not been able to find easterners in either Leadville or Denver willing to join the venture. Finally, despairing that his property would ever be developed, Hyman turned to a seasoned mine promoter, Thomas L. Ewing, Jr., a general during the Civil War and in 1880 congressman from Ohio, who was thought to be the representative of "large capitalists and had a big reputation as a mining expert and also as a reliable man." Under the terms of this new arrangement, Hyman agreed to pay Ewing $1,000 a month to develop

13 Hyman, "Mining Venture," pp. 12–13; Hall, *History*, 4: 273.
14 Hyman, "Mining Venture," pp. 15–16.

the claims, and, if they proved valuable, the congressman would promote them in the East and secure all the funds necessary to proceed
with full scale operations. In very short time, however, Ewing decided
that the property was not worth his attention and withdrew abruptly,
leaving the Cincinnatian to his own resources once again. Subsequent
attempts to enlist capital met with equally poor results. For a while it
appeared that Frederick Butterfield, a wealthy New York merchant,
whom Hyman believed had made millions of dollars in Leadville,
would save the scheme with a $150,000 investment, but Butterfield
died on a European vacation. Thus, by mid-1881 the Cincinnati attorney still held a controlling interest in the claims on Aspen Mountain
and to his chagrin had spent many times over his original "flier,"
without either selling the property at a profit or developing it into a
productive operation.[15]

Revealed in these experiences of David M. Hyman are several
basic characteristics of eastern involvement with the gold and silver
mines of Colorado: the gambling instinct, the reliance upon remote
agents, the need for large capital outlays, the intensity of promotional
efforts, and the role of friends and associates in starting and shaping
a mining venture. Depending, of course, on how fully they were
utilized, personal contacts and relationships furnished a natural network through which a highly speculative project in a distant territory
could draw capital investment and also permit the smooth function-
ing of a company when teamwork brought a high premium. Yet, on
the debit side of that same ledger must be accounted the problems of
kinfolk and cronies whose incompetence sabotaged sound operations.
One engineer noted that some companies became mere "asylums for
dependents" in which the blundering friends of eastern owners un-
ceremoniously ruined good mining properties. To guard itself against
the slightest hint of nepotism, the LaCrosse Mining, Milling, and
Power Company of Boulder vowed to create a profitable enterprise
under the full control of "practical" businessmen and with "no place
for sinecures, deadheads or idle and shiftless relatives to absorb the
profits of the undertaking." Although there was always room enough
for wiser, more economical management of the mines, personal asso-

[15] Ibid., pp. 17–35.

ciations formed essential links in the financial process and certainly made the investment market more responsive to mining projects.[16]

If he possessed the right friends and connections for financing his mine, according to one commentator, the prospector could have avoided falling into the clutches of that "conscienceless class" of professional promoters who finagled away his claim or at least asked commissions so high that profits on the sale never reached the discoverer. Former Confederate soldier turned prospector M. B. Shelton placed a hydraulic mining proposal—that is, one requiring a substantial sum of money to wash away the gold-bearing banks of a stream with powerful jets of water—in the East by initiating a promotion through friends. He first contacted a close friend and Mississippi-born lawyer in Georgetown, who agreed to send the plan to a judge he knew in the Chancery Court of his home state, who in turn accepted a principal part in funding the scheme and in all likelihood invited his neighbors in Friers Point, Mississippi, to join him as subscribers. Consequently the Leavenworth Boom Ditch Company resulted from these relationships formed outside the business of mining and depended upon personal advocacy that had reached an unexpected source of capital in the Deep South.[17]

By relying on personal ties and a chain of promotion, one man's speculation in Colorado often eventually bound together his tried and true friends, relatives and neighbors in a mining syndicate. A principal factor in this development was the absence of a large and impersonal money market in the nineteenth century for volatile gold and silver mines, where distance, charlatanism, and erratic streaks of ore, to name a few, played hob with the industry. Faced with this set of circumstances, the leader of a new venture normally sought financial support among friends, hopefully drawing upon their good will towards him and placing his own reputation behind the plan as an added incentive and measure of security for those he contacted. In essence, having a friend aboard ship gave a little ballast on the stormy seas of mining finance. Norvin Green of the Western Union admitted

[16] New York *Bullion*, comp., *Bullion: Its Production and Use*, p. 37; LaCrosse Mining, Milling, and Power Company, *Prospectus*, p. 14.

[17] Thomas A. Rickard, ed., *The Economics of Mining*, p. 113; M. B. Shelton, *Rocky Mountain Adventures*, pp. 70–72.

to taking a "small interest" in a mine and accepting a trusteeship in it to "oblige a personal friend," even though he added quickly that "like all mining property, it is purely speculative, and may pay large profits, or may result in total loss of investment." The same power of personal persuasion seemed to be at work in the Pelican and Dives Mining Company, headed by Green and having offices in the Western Union Building in New York. The company's directorate included the executive officers of two parcel express companies and of three telephone and telegraph companies, and the assistant postmaster general of the United States, all of whom undoubtedly knew each other well before combining to mine a slope of Sherman Mountain near the town of Silver Plume.[18]

By extending into the West the peculiar ways of working together that they had developed through regular contact in the East, men associated in the same firm or trade attempted to cooperate in a Colorado mining scheme and conduct its business along with other matters they handled jointly. Middle-class merchants, bankers, lawyers, and small manufacturers in many towns and cities across the country pooled their capital and, diverting it from local investments, took a flier in the western mines. The chief financiers of a large mining operation near Crested Butte in the early 1880s consisted of the assistant general manager, the freight agent, the land commissioner, and the cashier of the Santa Fe Railroad at Topeka, Kansas, plus two of the town's bankers. Though noted for their golden beverage rather than their silver ore, owners of two of the largest breweries in the United States, William J. Lemp and Adolphus Busch of St. Louis, put aside their competition to buy the Forest silver mine at Ouray. A New York bookseller and a publisher in that city helped to organize a mining company that owned property in Gilpin and Clear Creek counties. And four former generals in the Union Army—Edward M. McCook, Henry W. Slocum, James McQuade, and Adelbert Ames— commanded a large block of stock in the Snowdrift Consolidated Mining Company at Georgetown in 1880.[19]

[18] Green to W. H. Smith, September 10, 1879, and Green to Theodore N. Vail, April 21, 1880, Letterbrook 1, Green Papers; *Pelican and Dives Mining Company*, pp. 1–2.

[19] *EMJ* 28 (August 2, 1879): viii; *EMJ* 29 (April 24, 1880): 298; *EMJ* 45 (March 3, 1888): 167; *Mining Record* 7 (February 21, 1880, and May 8, 1880): 179, 456.

No mere coincidence will explain the affiliation of Henry B. Hyde and General Louis A. Fitzgerald in a handful of Colorado mining enterprises. Fitzgerald, another Civil War officer, acted as secretary-treasurer of the ill-fated Maid of the Mist Silver Mining Company in which Hyde owned a large quantity of stock, but it was only one of the many business interests the two men held in common. The general, as president of the Mercantile Trust Company, handled banking matters for Hyde's Equitable Life Assurance Society, which, in turn, leased office space to the bank, and both men held shares in the firm directed by the other. Together with a warm personal friendship, this commercial association facilitated the investments made by the two New Yorkers in mining as well as in other projects. On occasion, Hyde called upon Henry R. Wolcott of the Boston and Colorado Smelting Company for advice in mining matters, usually in regard to properties in the Rocky Mountains, and the original contact also seems to have been made outside the sphere of mining. Wolcott had negotiated with the Equitable on loans for a variety of business projects and he had been instrumental in Hyde's decision to construct an Equitable Building in Denver. Apparently the insurance company president knew Wolcott well enough to ask him to find a job in Colorado, and in mining for instance, for his twenty-two-year-old cousin who had left Yale University prematurely and who, said Hyde with characteristic bluntness, was in "perfect health but his head troubles him during his studies."[20]

The elastic code of ethics of American politics in the last third of the nineteenth century often permitted city halls, statehouses, and the federal Congress itself to becomes hives of business activity. One scholar found that "at times the [United States] Senate resembled the floor of a brokerage house" as senators unabashedly negotiated business deals and recommended investments to each other. It did not take long for some state legislators, mayors, governors, and a multitude of officials at all levels of government to discover that one perquisite of public office was the chance to make friends and contacts who might as readily be business partners as political ones. In Washington, D.C., businessmen-politicians of both major parties vigorously bought and

[20] Hyde to Fitzgerald, March 11, 1880, February 12, 1883, and Hyde to Wolcott, June 16, 1890, October 22, 1892, April 25, 1893, November 28, 1894, Hyde Papers.

sold real estate, traded railroad securities, and speculated in everything from coal to gold, continuing financial activities begun before entering office and partially explaining why the Senate alone boasted twenty-five multimillionaires at the turn of the century. Frequently, uniting on the basis of close personal and political ties, a large number of prominent national legislators took time from matters of state to pool their resources in a Colorado gold or silver mine and to help finance the western frontier.[21]

The organization of the Allied Mines Company in New York in the fall of 1880 with a capital stock of $10 million strikingly illustrated this activity. Sitting on its board of directors were Congressman Thomas L. Ewing, Jr., of Ohio, Senator Preston B. Plumb of Kansas, and former senator and Secretary of the Interior Orville H. Browning of Illinois, and listed among the principal stockholders in the enterprise were former senator Aaron H. Cragin and Senator Henry W. Blair of New Hampshire and Senator—soon to be Secretary of State —James G. Blaine of Maine; all were Republicans except Ewing, who had become a leading spokesman for the Greenback wing of the Democratic Party and a vocal advocate of silver remonetization. While there can be no question that these politicians consorted with each other in the halls of Congress and the party caucuses, a few had displayed close friendships prior to 1880, particularly with the central figure in the Allied enterprise, Thomas Ewing, Jr. For instance, the Ohio congressman and Preston Plumb had been leaders of the free state movement in "Bleeding Kansas" of the 1850s, both had been officers in the Eleventh Kansas Volunteer Infantry, and Plumb had been General Ewing's chief-of-staff and his provost for a time during the Civil War. And due in large measure to the relationship of Ewing's father with Browning as a law partner and as a fellow advisor to President Andrew Johnson, Ewing and the Illinois politician had developed a strong personal rapport. Furthermore, Ewing and Blaine were second cousins and had been frequent companions in their youth.[22]

[21] David J. Rothman, *Politics and Power: The United States Senate, 1869–1901*, pp. 208–09.

[22] David S. Muzzey, *James G. Blaine: A Political Idol of Other Days*, p. 13; *Allied Mines of Imogene Basin, Ouray County, Colorado*, pp. 1–3; *DAB*, s.v.

The Allied company's mining property proved quite as interesting as its owners, who reportedly paid $50,000 for some seventy acres
of gold- and silver-bearing ground on Mount Sneffles, a few miles
from Ouray. The claims had been discovered, promoted, and ultimately managed by William Weston, an Englishman with a certificate
in assaying from the Royal School of Mines who asserted that the
"true fissure veins" in the property would yield a net profit of "at
least" $60,000 a month. The claims never fulfilled these grandiose
expectations, mostly because of the expense in working in their isolated and perilous position on the mountainside, and so were subsequently abandoned by the politicians. However, in 1895 a prospector,
prowling in one of the derelict shafts, discovered that the development work had ceased just short of opening an enormous quantity of
telluride gold ore that ran as high as $3,000 of metal to the ton.
Thomas Walsh bought the property for $20,000, renamed it Camp
Bird, and recovered over $2.5 million in gold, enough to buy his wife
the celebrated Hope diamond. Nor did that exhaust the mine; between 1902 and 1918, new owners extracted another $18 million.[23]

Contacts with the crowd of congressional speculators in mining,
which at times represented a who's who of the national legislature,
became an invaluable asset, especially in the hands of a professional
promoter and operator of the caliber of Stephen W. Dorsey. A shady
and mischievous character, he helped to place several Colorado mines
with political and business friends in the 1880s and 1890s, despite
a public career that ought to have given pause to any but the most
gullible. After an uneventful tour of duty in the Union Army, Dorsey
had moved to Arkansas in 1868, where he presided over the Arkansas
Central Railway Company, a project that existed almost entirely and
fraudulently on state, county, and city financial aid aimed at encouraging legitimate railroad enterprises in the postwar era. Partially to
protect these interests, he entered politics and in 1873 had been
elected as a Republican to the United States Senate. He used this
position primarily as a stepping-stone into the upper echelons of his

"Blair, Henry," "Browning, Orville," "Ewing, Thomas," "Plumb, Preston"; *National Cyclopedia of American Biography*, s.v. "Cragin, Aaron."
[23] *Allied Mines*, pp. 5–11; Hall, *History*, 4: 250, 604–06.

party, and a good deal of his energy was spent in directing the national party campaign in 1876 and again in 1880. However, shortly after managing the successful presidential campaign of James A. Garfield in 1880, the Arkansas senator was indicted for conspiring to defraud the federal government of nearly $500,000 in the notorious "star route" affair, and, although acquitted, he was later charged with having attempted to bribe the jurors. His political fortunes shattered, for the next twenty years Dorsey chased speculative rainbows in the West, concentrating on cattle operations in the Southwest and the precious metals of New Mexico and Colorado.[24]

In Colorado, Dorsey established a reputation as a man with important political and business contacts, which he used to promote a variety of mining ventures. One malodorous scheme at Silver Cliff called the Bull-Domingo Mining Company united Dorsey and another senator and railroad man from Arkansas, Alexander McDonald, along with an agent for the Vanderbilts, the lieutenant-governor of New York, Connecticut senator William H. Barnum, and the renowned legal reformer and codifier David Dudley Field. For awhile the company prospered, but the Bull-Domingo eventually failed, owing to poor management and, in the view of some shareholders, to stock manipulation by "insiders." Dorsey turned his attention to the Leadville mines in the 1880s, and during the Cripple Creek boom of the 1890s he invested in a few properties as well as promoted several ventures for the partnership of David H. Moffat and Eben Smith, demonstrating once again his influence among men of wealth. In Detroit, for instance, he sold a large block of stock in the Victor gold mine to "Match King" Russell A. Alger, Michigan senator and a commander of the Grand Army of the Republic. Several years later, however, Dorsey reportedly fell into "his old habits" of getting drunk for days at a time and could no longer attend to his many business interests.[25]

The natural tendency of mine organizers to reach out to their friends and relatives for financial backing certainly did not mean that

[24] *DAB*, s.v. "Dorsey, Stephen"; *National Cyclopedia of American Biography,* s.v. "Dorsey, Stephen."

[25] *EMJ* 28 (September 13, 1879 and November 15, 1879): 190, 363; *EMJ* 29 (April 3, 1880): 250. Dorsey to Smith, May 9, 1894, Box 1; Jerome D. Gillett to Moffat, February 6, 1897, and F. L. Roudebush to Smith, February 6, 1897, Box 2; all in Eben Smith Papers, Denver Public Library.

the practice always enjoyed smooth sailing. On the contrary, an ordinarily sensitive man might have felt considerable discomfort in coaxing his associates into buying a piece of a highly speculative enterprise, and he might also have feared the ill effects a failure would have on all of his personal and business relationships. As a result, in some cases the responsibility weighed heavily on the mind of a businessman and raised serious questions about the ethical thing to do. Norvin Green faced such a problem in 1888 when the promoters of his United Claims Mining Company sent a circular to all Western Union managers suggesting that employees purchase a few shares of the stock. Green objected primarily because the pamphlet would fall into the hands of the "more ignorant" employees, who were "mostly" females on modest salaries, and the ones who could least afford an investment in anything so unsure as a mining company. Although he sincerely wanted to avoid that mistake, as president of the United Claims, he felt that he could not do so without it being "construed that I have no faith in the value of the property," and so "my position is rendered somewhat embarrassing." In an apparent compromise between his conscience and fresh capital from his employees, Green decided that the pamphlet must reach only the "higher grades of officials, who are perfectly competent to judge for themselves, and some of whom have money enough to take speculative chances." [26]

To be sure, when a mining operation ran into hard times and started to cost more money than it made, the strings of friendship and good will, once thought to be secure, began to stretch and snap. And especially if the promotion had been aggressive and full of lofty promises, the initial reactions of shock and disappointment could easily swell into a sense of personal betrayal. L. Bradford Prince, a descendent of Governor William Bradford of the Plymouth Colony, took advantage of his sterling reputation as a funds raiser and president of the Association of Church Chancellors in the Episcopal Church to sell stock in his mining schemes to bishops, clergymen, and other members of the denomination. Until his company near Idaho Springs suffered a series of setbacks, trust in Prince as a good man ran high, and then stockholder faith in him began to collapse. A pious

[26] The United Claims property was located in Chaffee County. *EMJ* 40 (December 12, 1885): 406; Green to W. G. Robinson and Company, August 15, 1888, Letterbook 4, Green Papers.

woman investor in Bunting Hollow, New York, prayed for his soul and hoped that "He whose is the whole earth, and fulness thereof" would somehow rescue the firm from bankruptcy, while a minister indignantly protested that "churchmen should have a care as to how they use prestige to induce the credulous to invest. It was on the reputation which the officers had gained that I induced my friends to take the stock in the first place." One of Prince's major stockholders, a seventy-nine-year-old New York businessman who owned property in the scenic Thousand Islands near Niagara Falls, had also promoted the mine, which he labeled "missionary work," by treating his business associates to a tour of the islands and asking all aboard the launch to take ten or twenty shares each as a "souvenir of their old companion." He too felt the anger of "friends" and passed on the blame to Prince for "this unfortunate enterprise."[27]

A similarly unpleasant feud between Cyrus McCormick and fellow Chicagoan John V. Farwell, a wealthy merchant and part owner of the sprawling XIT Ranch in the Texas Panhandle, erupted over a miscarried investment at Leadville and led to a law suit. Through an involved process, McCormick had joined Farwell and a group of eastern financiers in 1879–1880 to purchase the Little Chief mine that enjoyed a brief life before stock manipulation, poor management, and declining ore values caused it to falter, oddly enough, right beneath the nose of the company president, McCormick himself. Apparently the humor in this turn of events escaped the manufacturer, for in 1882 he brought suit against Farwell and others in the supposed inner circle of the company for losses in excess of $90,000. Caught by surprise when he received a court summons from his partner, Farwell feared above all else the adverse publicity that a trial was sure to create and, obviously unnerved, wrote McCormick, "I don't want any of my family to know that any such claim has been made." Despite his pleas that the matter be settled out of court, preferably by three mutual friends or "brethren" from the Presbyterian church both men attended, McCormick's son paid a visit to the merchant and rapidly "disabused his mind of any personal relation which might exist." Probably much to Farwell's discomfort, the suit dragged on, eventual-

[27] DAB, s.v. "Prince, LeBaron"; S. G. Pope to Prince, May 13, 1895, September 9, 1899; Mary Dickinson to Prince, August 18, 1897; Reverend C. L. Fulforth to Prince, undated; all in Prince Papers.

ly reaching the New York State Supreme Court, though no verdict was reached before McCormick died in 1884.[28]

An inescapable corollary of personal finance was the association of one small town, city, or section with a particular mining enterprise in the West. Normally taking the form of a moderately capitalized company, and usually giving local businessmen their first experience with an investment far from home, such ventures may have warmed the pages of the hometown newspaper for a while and turned barbershop conversations to gold lodes and silver streaks for a spell but stirred little interest elsewhere. In the Colorado mining districts, however, visitors saw abundant evidence of them, such as a Pittsburgh and San Juan, or a Kansas City and Aspen mining company, or the New England and Cripple Creek Gold Mining Company, which had been organized by people in the Boston suburbs of Somerville, Concord, and Malden. Ownership of the Lulu mine in Clear Creek County rested in the hands of Peoria County, Illinois, men: a banker, a merchant, a sheriff and his deputy, the county treasurer, and the clerk of the circuit court. In a similar way the Colorado and Galveston Mining Company at Silver Cliff represented the cooperation of local tradesmen, bankers, and other small businessmen in that East Texas cotton town. And in 1881, several prominent citizens in the prairie towns of Champaign-Urbana, Illinois, formed the Poverty Gulch Mining Company, with a capital stock of one million dollars, to work seven claims near Crested Butte. The chief promoter was a dentist who seems to have extracted stock subscriptions, among other things, from his patients, one of whom, an Urbana bank president, became heavily involved and presumably drafted his friends and customers into the speculation. Local newspapers raved about the richness of the mines and communicated an air of excitement to the whole community. Unfortunately, however, Poverty Gulch proved to be a fair description of the whole project, and in 1884–1885 the small syndicate sold its Colorado property and retreated from the business of mining.[29]

---

[28] Charles C. Copeland to McCormick, May 22, 1882; Farwell to McCormick, October 20, 1882, November 30, 1882, December 25, 1882; Cyrus H. McCormick, Jr., to McCormick, October 25, 1882; all in Series 2A, Box 34, Cyrus H. McCormick Papers, State Historical Society of Wisconsin, Madison. William T. Hutchinson, *Cyrus Hall McCormick*, 2: 199.

[29] *EMJ* 28 (December 13, 1879): 438; *EMJ* 41 (July 17, 1886): 48; *EMJ*

Once partners in a mining venture, some investors tried mightily to make it work and succeed, rather than trusting their capital to the whims of fortune or resigning themselves to a certain loss when trouble hit the project. Consequently, what might have begun as a gamble involving a relatively small amount of money evolved into a more serious and demanding proposition than first expected and one that challenged the skill and determination of its owner. To protect his own interests, it behooved an easterner to read the mining news columns, keep track of changes in the industry, study the reports of his manager in the field, and generally assume some of the same responsibilities toward this investment that his other business and professional affairs required. In the process of devoting more and more time and energy to the mines, he may have slowly slipped away from his previous occupation and increasingly staked his future on the gold and silver deposits of the West. On the whole, this transformation was more likely to occur among men with a solid business background than among the adventurers who gathered daily at the stock exchanges to dabble in every passing speculation.

Because of his investment at Aspen, David Hyman increasingly neglected his law practice in Cincinnati and let other matters at home slide, even though he "never contemplated that under any circumstance that I would abandon the bar for I was much attached to my profession and I had no reason whatsoever for changing my business career." The troubles he encountered in purchasing the mine and then protecting it from claim jumpers in the area meant the expenditure of more money than he ever imagined, a good deal of it going for the services of lawyers and mining engineers, and several long and difficult trips westward to inspect the mine and direct its development. To make matters worse, his determination to proceed with the Aspen enterprise, which in the early stages looked to be an exercise in futility, harmed his business in Cincinnati, and when at home he found himself "short of clients and short of money and much depressed" due to his Colorado activities. Hyman's efforts finally began to pay handsomely in 1884–1885 when his miners struck high-grade ore,

66 (December 31, 1898): 796; *Mining Record* 8 (July 31, 1880): 109; Paul S. Barnett, "Colorado Domestic Business Corporations, 1859–1900" (Ph.D. diss., University of Illinois, 1966), pp. 199–202.

and ultimately the property produced millions of dollars in silver. With that, the practice of law came to an end, and for the next thirty years Hyman concentrated on expanding his business affairs in the West, saying apologetically in his old age that once involved with mining, he could never bear to cut "completely loose" from it.[30]

Similar feelings partially explained the more than ten years that Joseph Reynolds of Chicago devoted to western mining after a long and very successful career as a packet-line operator on the Mississippi River and as a railroad builder. Known popularly as "Diamond Joe" from the trademark on his sternwheelers that carried grain between St. Louis and St. Paul, the millionaire shipper purchased mining property in Leadville, in Pitkin County, and on Red Elephant Mountain near Lawson, as well as in Arizona. Frequent trips to the West and closer attention to his investments became necessary when Reynolds discovered that one of his mines had been salted and that the manager of his Crown Point mine at Leadville, though reputedly unable to read or write, had filched about $250,000 out of the ore receipts. Reynolds tried to prevent a recurrence of these frauds and moved west to direct his operations personally. At the age of seventy-two, he died in a shack near Prescott, Arizona, while developing one of his gold mines.[31]

Between periodic blasts at the dangers of mining speculation, Norvin Green spent considerable time in his New York office trying to follow the number of mine speculations that had captured him. Green admitted to making his first investment in 1879 when he put $2,000 into a Colorado property promoted by a Burlington, Vermont, toboggan builder. Ten years later, the president of Western Union wrote the promoter complaining that he had never received a profit on his money and took the opportunity to give a brief history of his association with mining and the causes of his present distress; until 1879 Green had strongly "resisted" any proposal, but since then he had invested in Colorado, Nevada, and Mexico, had spent more than $100,000, apparently sending good money after bad in each case, and

[30] Hyman, "Mining Venture," pp. 11–17, 23–28, 121.

[31] *EMJ* 25 (March 16, 1878): 188; *EMJ* 26 (December 16, 1878): 206; *EMJ* 39 (May 30, 1885): 377; Thomas W. Goodspeed, "Joseph Reynolds," *The University Record* (University of Chicago) 7 (January, 1921): 31–38; Additional biographical information is in William J. Peterson, "Joseph Reynolds," *The Palimpsest* 51 (April, 1970): 169–78.

could report no profits from any of these transactions. "My courage in mining investments," he said, "has pretty much oozed out at the fingers' ends, and I am going to go very slow." With substantial interests in several mines and the presidency of two mining companies, Green could not break away suddenly, and, a few months before his death in 1893, he busied himself soliciting funds from other capitalists to save his United Claims enterprise in Colorado. Win, lose, or draw, Green had shown, as did Hyman and Reynolds to a different degree, that mining fever could also afflict the more experienced and conservative elements in American society at the end of the century. Furthermore, through their persistent efforts to make money from the mines, they added a measure of stability to an impetuous industry.[32]

Firmly held ideas about mining as merely a form of gambling did not completely divert some investors and their advisers from serious consideration of the risk factor. How certain a mine owner could be that his property would return a profit was a question deliberated by engineers, journalists, and, of course, capitalists, though always with inconclusive results. Basically, since ore did not occur in any uniform pattern and could peter out suddenly, most experts granted that fate held an option in every gold and silver mine. As T. A. Rickard observed, the fellow who hoped to remove this element entirely no doubt also expected "to go bathing without getting wet." Still, the potential investor had some choices left to him within the broad limits of this riskiness; he could at least judge a property by its stage of development, which could range from an unopened claim to an extensively equipped and proved mine that had a history of paying dividends. Typically, the more knowledgeable eastern investors preferred a property that gave some credible evidence of its value, but one that held out the promise of yielding the greatest, if not also the surest, amount of profit at the least possible cost.[33]

Advice on this score appeared from both sides, those endorsing largely unproved claims and those urging well-developed mines in which the ore content had been carefully measured and blocked out

[32] Green to Wilbur, Jackson and Company, July 1, 1885, Letterbook 3; Green to Charles H. Emerson, January 29, 1889, Letterbook 4; Green to Edward A. Server, January 9, 1890, Letterbook 5; Green to W. J. Van Patten, November 4, 1892, Letterbook 6; all in Green Papers.

[33] Rickard, ed., *Economics of Mining*, p. 69.

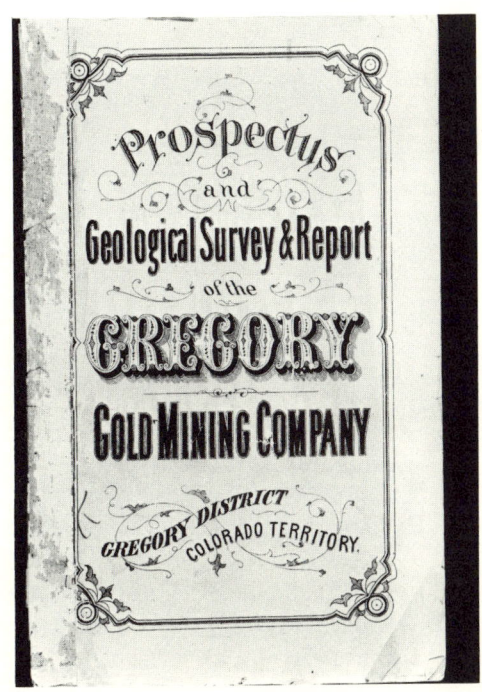

Typical mining company prospectus of the 1860s. *Denver Public Library, Western History Department.*

An artist's view of New Yorkers lining up to buy mining stock. *From William W. Fowler, Twenty Years of Inside Life in Wall Street (1880), Greenwood Press reprint, 1968.*

LEFT: Bostonian J. Parker Whitney promoted the Colorado mines for thirty years. *The State Historical Society of Colorado.* RIGHT: George D. Roberts introduced "California methods" of stock promotion in the East. *California Section, California State Library.*

"Fitz-John's Folly." Stamp mill of the Gunnell Gold Mining Company, Black Hawk, 1864. *The State Historical Society of Colorado.*

Little Chief Mine, Leadville. *The State Historical Society of Colorado.*

"A Wall Street Man's Experiences in Leadville." Cartoon in *Harper's Monthly* (February 1880) depicts arrival of an eastern speculator in Leadville. *The State Historical Society of Colorado.*

View of the Chrysolite Silver Mine, Leadville. *The State Historical Society of Colorado.*

Hauling heavy machinery for the mines. *The State Historical Society of Colorado.*

Drilling with single and double jacks. *The State Historical Society of Colorado.*

LEFT: David H. Moffat, Colorado banker and railroad builder, owned mining properties throughout the state. RIGHT: Eben Smith, able mine manager and partner of David H. Moffat. *The State Historical Society of Colorado.*

Mining operations at Gold Hoist Mine, Cripple Creek. *Denver Public Library, Western History Department.*

LEFT: Norvin Green, president of Western Union, doubted the worth of mining investments. *Warren T. Green.* RIGHT: Henry B. Hyde, organizer of the Equitable Life Assurance Society, reported large losses from his mining interests. *Baker Library, Harvard University.*

Cyrus McCormick, inventor and manufacturer, invested at Leadville. *Painting by Alexander Cabanel, 1867. State Historical Society of Wisconsin.*

Bidding for shares on the floor of the Cripple Creek Mining Exchange around 1900. *Denver Public Library, Western History Department.*

LEFT: David M. Hyman, principal figure in costly mining litigation at Aspen. *Aspen Historical Society*. RIGHT: Joseph ("Diamond Jo") Reynolds turned to mining after developing a successful steamboat line on the Mississippi. *Special Collections, University of Chicago Library*.

by trained engineers. Yet, a few experienced mining men expressed greater concern that no matter what the absentee capitalist bought, it ought to be a mine and not simply a "gopher hole" foisted by a charlatan. From the *Mining and Scientific Press* came the opinion that, after "judicious selection," a financier was wiser to take "slight chances on the loss of a small amount [of money] in the development of virgin claims than the greater loss on old, exhausted or nearly exhausted dividend mines." While that view starkly contrasted virgin claims with nearly exhausted mines, most properties actually fell somewhere in between the two and were ordinarily described as being "partially developed" by shrewd promoters who wanted ample room to negotiate with buyers. On the whole, the closing stages in a good mine's life were less likely to produce high returns on an investment than the same mine would, had it been purchased in its youth at a naturally lower cost. Yet, an older mine with its ore value known permitted a "safer" investment even though the profit margin tended to be slimmer. One New York operator unhesitatingly recommended undeveloped ground, since the profits, according to his reckoning, would be 50 or 100 percent a year, or as high as 400 or 500 percent, leading him to conclude that it made a "better speculation."[34]

In practice, both new and established mines attracted the larger capitalists. Isaac L. Ellwood, for example, followed the lead of his engineer, John B. Farish, in buying little more than an outcrop of silver ore near Ouray in 1894. Under Farish's supervision and with Ellswood's cash, the claim advanced from a hole in the ground served by a crude horse-drawn whim, to a deep shaft worked by a second-hand hoisting machine, and finally, after many months, into a fully equipped mine capable of producing large and steady profits. Norvin Green seems to have spread his money over undeveloped and developed locations. In the latter category, he headed the Pelican and Dives Mining Company, whose property at Georgetown had been worked extensively. However, due to ten years of litigation and feuding by a succession of owners and claimants, the mine had been badly abused, in part, by frenzied efforts to gouge from it the highest grade ore in order to hire a legion of lawyers. Consequently, when Green's

[34] Henry B. Clifford, *Rocks in the Road to Fortune or the Unsound Side of Mining*, p. 22; *MSP* 43 (September 17, 1881): 188; J. S. Buell, *Suggestions Relating to the Organization of Gold and Silver Mining Corporations*, p. 7.

company took control in 1880, it estimated that $50,000 would be needed for repairs, opening ore reserves, and constructing a mill, yet the $3 million in silver reportedly produced in earlier years gave Green confidence that the enterprise was surer than most and that it could pay dividends almost at once. Apparently after due consideration of the risks, hard-headed men of business such as Ellwood and Green had reached different conclusions about the best kind of mining speculation in Colorado, and both capitalists no doubt believed that they could be prudent and daring at the same time.[35]

Of course, even under the most favorable circumstances, no "sure things" for eastern investors cropped up in the western mines. The pendulous swings of the frontier industry between boom and bust, wild speculation and costly development work, could readily account for huge profits or vexatious losses, with history inevitably remembering the winners and forgetting the losers. While the extravagant success of Meyer Guggenheim in Colorado has been told and retold, scant attention has been given to the bad investments of Norvin Green that consumed thousands of his dollars or the several enterprises of Cyrus McCormick that, at best, balanced profit with loss. Good fortune in the Colorado gold and silver mines naturally had its own generous rewards, but the less fortunate had to take comfort in their hard-earned lessons about the uncertainty of mining or, with Henry Hyde, smile at their own bad luck and invest no more. Hyde, whose investments had been superbly unproductive, bantered with a promoter from Colorado Springs:

I have had bad luck with gold mines so far and I should be exceedingly sorry to break up all the coming prosperity which is in store for you. I am as certain as I am of anything on earth that if I owned one share of the thing you speak of, it wouldn't be worth a cent in less than a year. I have ruined at least a dozen of my friends by buying a few shares of stock in companies which they represented. I have arrived at that time of life when my wants are few—peace, tranquility and a little western air are all that I need. I do not know but that by and by I shall apply for a position in the Mammon gold mine for the purpose of re-

[35] Farish to Ellwood, December 31, 1894 (Box 26), January 4, 1894 (Box 27), December 2, 1896 (Box 37), Isaac L. Ellwood Papers, Western History Research Center, University of Wyoming, Laramie; Fossett, *Gold and Silver Mines,* pp. 391–93; *Pelican and Dives,* pp. 3, 13–14.

newing my health and strength as I am about tired out in my work in the east . . . The fact of it is, I am afraid it would have a bad effect upon my character if I should make a great deal of money in any new venture.

Beneath the humor at his own expense, Hyde expressed the resignation of absentee capitalists who had put more money into mining than they had taken out. For Chauncey Depew of the New York Central Railroad, twenty-six mining ventures in which he "lost every dollar" had filled him with such "horror" that he distrusted gold or silver "even after it is assayed and had the stamp of the United States Mint upon it."[36]

Still, the pleasant thought of removing millions of dollars in precious metal from the Colorado mountains could not be repressed. If the mines seemed to be the quickest way to a vast fortune, then bigger and better opportunities awaited the capitalist who unlocked his purse, seized the main chance, and sent his dollars west to the magnificent silver fields around Leadville.

[36] Hyde to W. R. Varker, May 14, 1892, Hyde Papers; Chauncey M. Depew to Prince, November 8, 1894, Prince Papers.

4

East of Leadville

A s the decade of the 1870s reached midpoint, there was marked
uncertainty in Colorado about the future of the hardrock mining industry. Capital investment from the East had dropped off sharply in
the wake of the Panic of 1873, which disrupted nearly all business
activities and discouraged even the most bullish investor from plung-
ing into the briar patch of western mining. Early in 1876 the full
effects of this crisis in the national economy were being felt in Colorado, with work on dozens of mines in the initial stages of development being halted and many more good prospects being left unopened and untested.

Gone were the giddy days, thought some Coloradans, when
bonanza ore deposits and flush times in the East could combine to
ignite mammoth mining booms that brought forth an avalanche of
absentee capital. Certainly easterners would recall with a cringe those
swollen dreams and millions of greenbacks interred beneath the hills
and gullies of Central City and Black Hawk between 1863 and 1865.
And for outsiders, Colorado evoked an image of worthless properties,
reduction hoaxes, and a graveyard for New York and Boston money
that would not be easily erased. Doubts and grim predictions about
the future spread through the mining towns in 1876 and 1877 and
cast a long shadow over the excitement of Colorado's first years of
statehood.[1]

When a pair of seasoned prospectors, William H. Stevens and
A. B. Wood from the Michigan copper fields, discovered in 1875
that the heavy, black sand clogging their sluices in California Gulch

---

[1] *EMJ* 19 (January 9, 1875): 21–22; *EMJ* 23 (May 26, 1877): 353; Frank
Hall, *History of the State of Colorado*, 2: 290.

were carbonates of lead rich in high-grade silver ore, eastern reactions were predictably slow. Deep-level mining required vast sums of money, which capitalists were understandably reluctant to commit at a time of economic contraction when surer investment possibilities than mining appealed for their financial support. So in spite of the ambitious ring to the Oro Mining Ditch and Fluming Company, the enterprise formed by Stevens and Wood rested principally on their own labor and the modest profits made from years of working the surface gold deposits in the upper Arkansas River Valley. Though the partnership staked several claims on what came to be called Iron Hill on the north bank of California Gulch, large-scale development awaited outside capital. Meanwhile, Stevens reputedly sold everything he owned, most of his wife's property, and borrowed whenever and wherever he could to purchase supplies and maintain a small work force on the claims. More than a year after his discovery, Stevens still had been unable to convince old friends and acquaintances around Detroit that he possessed valuable silver properties or that financially the times were ripe for a mining venture.[2]

Not until late in 1876 and early in 1877 did signs of a break appear in the taut money market. Eastern capitalists, while decidedly in favor of safer, longer-term investment outlets, began to show a renewed confidence in the economy, indicating that the darkest days of the depression had passed. For the camp in California Gulch, hopes for the future buoyed in 1877 when the wealthy Chicago merchant Levi Leiter purchaser Wood's half of the Iron Hill claims for an estimated $40,000. Bolstering that good news came word that August R. Meyer, a highly skilled metallurgist in the employ of the St. Louis Smelting and Refining Company, had recommended the erection of a smelter to treat the ores from the carbonate properties, which, due to their lead content, had to be smelted. Meyer and his employer, Edwin Harrison of St. Louis, began building a water-jacket furnace in May of 1877, contracted to receive ore, and ordered the construction of roads between the new smelter and the mines and the railroads pushing towards the Arkansas Valley. By displaying confi-

[2] Lewis C. Gandy, *The Tabors: A Footnote of Western History*, pp. 168–77; Don L. and Jean H. Griswold, *The Carbonate Camp Called Leadville*, pp. 172–73; Frank Fossett, *Colorado, Its Gold and Silved Mines, Farms and Stock Ranges, and Health and Pleasure Resorts*, pp. 408–10.

dence in the value of the area's mineral resources, these were important initiatives; they helped pave the way for further eastern investment as well as encouraging greater exploration of the surrounding
countryside.[3]

In a short time the scarred and nearly exhausted gold fields at
the head of the Arkansas swarmed with prospectors determined to
find the new treasure of silver. They fanned out in all directions.
Some inevitably stayed close to the operations of Stevens and Leiter,
while others combed the adjoining hills and ravines for glimpses of
metalliferous rock and a place to sink their shallow-prospect holes.
The strike everyone had hoped to make was made in the spring of
1878 by a pair of wandering miners named August Rische and George
Hook, who had recently been furnished with a few dollars in supplies, a grubstake in the miners' jargon, by Horace A. W. Tabor, an
Oro City grocer and soon to be millionaire. With the usual celerity
among poor men in a new mining camp, word of the discovery
spread, and before long a multitude of fresh claims surrounded this
Little Pittsburg location on Fryer Hill. Tabor, an uncomplicated
man whose burning desire to wrest a fortune from the mountains had
not dimmed since his arrival in Colorado in 1860, wisely recognized
the potential value of the property and set a gang of laborers to opening the ground. By early summer, 1878, his incline shaft had hit an
enormous deposit of rich ore, running as high as two hundred ounces
of silver to the ton and returning thousands of dollars in profit each
week. From all indications, the mine seemed certain to surpass every
other discovery in the state in richness. Fully aware of this possibility,
Tabor and Rische gratefully bought Hook's one-third interest for
$98,000 as the fall approached. Meantime, Leadville, as the upstart
town at the foot of Fryer Hill was called, quickly became the mecca
for an assortment of profit seekers.[4]

Without question the best-known and probably the richest vis-

[3] Gandy, *Tabors*, pp. 176–79; Hall, *History*, 2: 431; L. A. Kent, *Leadville.
The City. Mines and Bullion Product. Personal Histories of Prominent Citizens.
Facts and Figures Never Before Given to the Public*, pp. 179–80.

[4] Oro City, the original settlement in California Gulch, was located approximately one mile east of the present city of Leadville. Gandy, *Tabors*, 181–90; *EMJ*
26 (October 5, 1878): 243; G. Thomas Ingham, *Digging Gold Among the Rockies*,
pp. 399–400.

itor to Leadville that fall was Jerome B. Chaffee, United States senator from Colorado and president of the powerful First National Bank of Denver. As early as 1870, a reporter for the credit agency of R. G. Dun had estimated his wealth at six million dollars, placing him in the first rank of entrepreneurs who had grown up with the state, primarily through the mining business. Naturally welcomed by a deluge of mining proposals in the "Cloud City," Chaffee had come to examine the New Discovery claim, half of which he had taken an option upon soon after George Fryer made the location in the spring. Situated in Stray Horse Gulch and no more than seventy-five feet from the Little Pittsburgh, the New Discovery seemed to have the same brilliant future as its neighbor and to be clearly worth the $50,000 Chaffee now paid for Fryer's one-half interest. Before the senator could return to Denver, however, Tabor and Rische, knowing better than most what the carbonate beds could produce and dreading the possibility of litigation should they drift from their main shaft into the New Discovery, paid him $125,000 for his new acquisition, perhaps the quickest profit in Chaffee's life. Yet, the senator had second thoughts about this masterful trade a few days later. Drawing upon his "invaluable associate" in Denver, David H. Moffat, cashier of the First National Bank, Chaffee purchased Rische's interest in the New Discovery, Little Pittsburg, and adjoining claims for approximately $262,000. A short time later the properties were united and named the Little Pittsburg mine under the control of Tabor, Chaffee, and Moffat. As this series of transactions illustrates, the value of Fryer Hill acreage began to jump in a geometric progression and continued to spiral upward as miners unearthed new deposits of ore.[5]

Although the bonanza ore deposits had created a stampede of penniless prospectors and regional businessmen to Leadville, a full-scale mining boom, with a mad scramble by outside capitalists to share in the riches and the risks, had not yet been ignited. On that depended a revitalized economy, improved communications between

[5] Tabor's role in developing the Little Pittsburg is told best by Duane A. Smith, *Horace Tabor: His Life and the Legend*, pp. 71–75. Ingham, *Digging Gold*, pp. 402–03; Hall, *History*, 2: 435–46; Fossett, *Gold and Silver Mines*, p. 453; William B. Vickers, *History of the City of Denver, Arapahoe County, and Colorado*, p. 238; "Colorado," Vol. 1, Dun and Bradstreet Papers, Baker Library, Harvard University, Cambridge, Mass.

East and West, hard-driving promotion, and the public's conviction that the Fryer Hill mines offered spectacular profit margins. To a great extent the future of Leadville hung in the balance, since there were sharp limitations on how far local merchants or Denver bankers could go in meeting the financial needs of a field as large as that in California Gulch, even in these youthful stages of its growth. Purchasing a property represented only the first tentative step forward; ahead lay the high costs of labor, professional advice, transportation, machinery, and reduction facilities, which were essential elements in the production of ore but largely beyond the means of capital-short westerners. Like the pioneering investment of Levi Leiter, the mineral region required more eastern money in a steady flow to develop its productive capacity to greater depths.

Further relief from the business stagnation came in 1878 and spurred investor interest in the Fryer Hill discoveries, particularly during the last half of that year as businessmen seemed on the verge of a new burst of investment activity that would wipe away the depression doldrums. Probably influenced by the economic upswing and by his Chicago partner Leiter's success in the Iron mine, Marshall Field contributed toward the purchase of the Chrysolite claim and a few others near it. At the moment that he and his associate, Horace Tabor, took control of the Chrysolite, it was little more than the proverbial hole in the ground, reputedly "salted" to boot, and was deemed valuable because of its proximity to the Little Pittsburg. Yet the pair, along with a third owner, supplied the funds for development work that quickly unveiled a huge block of ore only a few feet beneath the surface. In December of 1878, another Chicago merchant and no newcomer to Colorado mining, John V. Farwell, led a syndicate of Illinois businessmen into the purchase of the Little Chief claim, sandwiched between the Chrysolite and the Little Pittsburg, at a cost of approximately $300,000. Almost at once, they began to sink several shafts, import steam-powered machinery, erect mine buildings, and plan the construction of a thirty-five-ton smelter on the site. Similar to Field and Tabor before them, Farwell and associates had taken over a promising, though untested, property that they intended to develop into a paying proposition, and, rather than rushing headlong into the formation of a stock company and floating shares on a still shaky public market, both groups steadily opened and improved their mines from their own financial reserves.[6]

Meanwhile, the formation of the Bullion Club in New York at the close of 1878 demonstrated the aroused curiosity of easterners in the western mines. The club was formed by a small group of interested brokers and promoters, apparently in hopes of establishing a market for mining ventures in between the conservative New York Stock Exchange and the disreputable private exchanges, called bucket shops, that were merely gambling dens. The group met in November in the office of Ross C. Stone, a dealer in mining securities, and organized a "social club" to stimulate investment in the West and prevent worthless proposals from reaching the New York market. According to its charter, the club would foster all legitimate mining operations (but no one in particular) and it would summarily suspend any member for "knowingly aiding or abetting in any manner any enterprise of questionable character." In less than two months, the Bullion Club boasted seventy-five members, mostly New York bankers and businessmen, who gathered each Thursday night in a hall in lower Manhattan to talk mining and to hear lectures from an imposing array of scientists and engineers. On any given evening, the membership might be treated to a pedagogical Benjamin Silliman, Jr., the dean of American chemists as well as a mining consultant, or an informed but spritely Rossiter W. Raymond, editor of the *Engineering and Mining Journal.* In part at least, the club attempted to cloak mining in a robe of respectability and gravity hitherto unknown, but at every meeting the postlecture conversation turned to the prospects of investing in the mines. Once the formal part of the program ended, all the elements in a mining venture circulated under the same roof: the latest reports on a new discovery in the West, professional mining engineers, scientists, experienced promoters, and, of course, some of New York's leading financiers.[7]

---

[6] For the trends in the American economy, see Rendigs Fëls, *American Business Cycles, 1865–1879,* pp. 83–124. Fossett, *Gold and Silver Mines,* pp. 445, 450; *EMJ* 28 (October 25, 1879): 301; Hall, *History,* 2: 439–40.

[7] The club had no president. The highest officials seem to have been the chairman of the board of trustees, a New York broker named McGinnis, and the secretary, David G. Croly, an editor and reformer. Croly's son Herbert strongly influenced

Quickly the Bullion Club became a catalytic agent in the mining revival. Over succeeding months the membership list grew and continued to broaden its spectrum of business leaders, including the publisher of an encyclopedia, the vice-president of the Pullman Palace Car Company, and Horace White, a nationally known political-economic reformer and former editor of the *Chicago Tribune*. If further evidence was needed that this would be no asylum for worn-out brokers, the club claimed regular correspondence with naturalist Alexander Agassiz of Harvard, Major John Wesley Powell of the United States Geological Survey, Senator Henry M. Teller of Colorado, and English Prime Minister William E. Gladstone, all of whom enjoyed honorary membership in the association. It is not surprising that by the winter of 1879 the club had reportedly become the "rage" in New York and that, in the view of some journalists, it had a unique opportunity to exercise a healthy and moderating influence on eastern investment. Ross Stone, apparently discontented with such an unrewarding role, sought to capture greater leadership in the excitement by issuing a bi-monthly journal called *Bullion* in March, 1879, as the semiofficial organ of the club. Unlike the organization, the journal would criticize or commend mineral properties to the buying public. For nearly two years, *Bullion* carried out this function vigorously, simultaneously reprinting lectures given at the Bullion Club and drawing upon the good name of that group, while editor and promoter Stone passed judgment on mining companies and securities.[8]

As the waves of a new mining boom were sweeping into New York and the East, the tide of fortune that had made San Francisco the queen city of the West and unrivaled center of mining finance seemed to be ebbing. For nearly two decades the city and its coterie of mining financiers and operators had prospered from a desolate section of western Nevada known as Washoe, where in 1859–1860 prospectors had discovered the Comstock Lode, a ridge of superbly rich silver ore about four miles in length. The speculative and exploitative ways of the "Bonanza Kings" had hollowed out approximately $300 million in silver from the mines around Virginia City

---

the progressive moment. *United States Annual Mining Review and Stock Ledger*, pp. 27–29; *Bullion* 1 (July 1, 1879): 2.

8 *Bullion* 1 (July 1, 1879): 2; *EMJ* 27 (February 1, 1879): 70.

and hurried the region into a premature old age. Now, with the bonanza days nearing an end, the future of the district would be left to its leaner ore deposits and the advent of new technology and stricter economies. For some of those San Franciscans who had benefited from years of rich returns and even richer speculations, the decline of the Comstock brought personal ruin, while for others—promoters, brokers, engineers, and, of course, the gamblers on Montgomery Street—it meant fewer opportunities and diminished fortunes. One prominent casualty of the downturn was John P. Jones, United States senator from Nevada and San Francisco mining millionaire, who, a few years earlier, had convincingly demonstrated his wealth by "buying" the town of Santa Monica as a Pacific seaport for one of his railroad schemes. In 1878, Jones reportedly suffered such great losses on the Comstock that he was forced to hock his wife's jewels and mortgage large tracts of choice real estate. Unlike the contrived drops in ore reduction that had characterized Comstock speculation in the past, in 1878 and 1879 the Nevada bonanza had actually begun to "peter out."[9]

To make matters worse, a new constitution adopted by California in 1879 showed the influence of a demagogic Irishman named Denis Kearney, leader of a workingmen's party, whose loud denunciations of privileged capitalists had been unsettling the San Francisco mining crowd since 1877. In the vacant lots around city hall Kearney stirred his listeners with calls for a few "judicious" hangings and similar vigilante appeals, aimed as much at intimidating the nabobs as galvanizing his own forces into legitimate political action. The Kearneyites he managed to place in the constitutional convention supported drastic reform of stock-market operations, and the dealers on Montgomery Street regarded such reform as an unbearable attack upon their interests. In particular, the chairman of the San Francisco Stock and Exchange Board termed "disastrous" a clause encouraging buyers to sue their brokers for losses sustained in stock-market

[9] The speculators of San Francisco and Virginia City received no sympathy from one mining review, which snapped that "there are probably more fools in these two cities than exist on the same amount of territory in any part of the habitable globe" (*United States Annual*, p. 19). George L. Upshur, *As I Recall Them: Memories of Crowded Years*, pp. 123–24; *Pacific Coast Annual Mining Review and Stock Ledger*, p. 27; Rodman Paul, *Mining Frontiers of the Far West, 1848–1880*, pp. 80–86.

transactions; another section equally abhorrent to the brokers prohibited margin sales. For the most part, however, mining operators
interpreted the new constitution as a further stage in the unchecked
"ruffianism" that had started in the sand lots and now seemed determined to drive the capitalists from the city through a combination of
restrictive laws and threats from a radical mob. Along with the falter-
ing Comstock mines, Kearneyism began to turn a number of mining
financiers and promoters away from Nevada and away from the
sagging fortunes of San Francisco to greener pastures.[10]

Just as the miners, mine managers, and engineers who had developed their skills in the Comstock Lode were leaving Nevada by
1878 for newer mineral regions in the West, the mining operators
of San Francisco sought fresh opportunities in Idaho, Montana, and
Colorado consistent with their particular talents and experience. Colorado, which had been nearly the exclusive preserve of eastern financing during the heyday of the Comstock, now suddenly became the
target of California curiosity, money, and, eventually, promotional
activities. In October of 1878, a correspondent for a New York mining journal noted the arrival of several California capitalists in Silver
Cliff, a recently opened camp in the Sangre de Cristo Mountains, and
wondered why these men were "getting more and more of a foothold
in this state," which he believed belonged "geographically" to eastern
entrepreneurs. His surprise probably grew in succeeding months as
many more groups of California promoters and financiers rushed into
the Rockies, slowed only for a moment by winter snows that choked
the mountain passes and made the newer, more remote fields virtually inaccessible. It appeared, however, that the excitement surrounding
Leadville would not be cooled by arctic temperatures or the hazards
of winter in the mountains, for the carbonate fields attracted a stream
of Pacific Coast visitors, reportedly ready to "carry off some of the
greatest prizes at Leadville."[11]

Among the Californians scouting Colorado in the spring and

[10] Kearnyism and the Constitution of 1879 are discussed by John W. Caughey,
California, pp. 386–91. Also, see Gerald D. Nash, "Government and Business: A
Case Study of State Regulation of Corporate Securities, 1850–1933," *Business History Review* 38 (Summer, 1964): 153–54. Joseph L. King, *History of the San Francisco Stock and Exchange Board*, p. 293; *New York Times*, February 21, 1880, p. 3.

[11] H. R. Vandemoer, comp., *Diaries, Newspaper Articles, and Letters of John
J. Vandemoer*, p. 157; *EMJ* 27 (February 15, 1879): 122.

summer of 1879 was George D. Roberts, pictured as a "small-sized individual, with a large head, a quick eye, and a pleasant smile," who had earned an enviable if not unsullied reputation as a shrewd stock operator in San Francisco. In addition to a long list of mining interests spread over California, Nevada, and New Mexico, he had been among the first to recognize the speculative possibilities in reclaiming the tule (swamp) lands of the San Joaquin and Sacramento valleys, though that project proved generally less rewarding than a few of his artfully managed mines. For a time in 1872, Roberts attained a brief bit of national notoriety for his part in the great diamond hoax in which South African diamonds were used to salt barren fields in northwestern Colorado. Despite this fiasco, his stature as an inventive promoter-speculator had been undiminished, and he maintained close ties with the principal mining adventurers of the Pacific Coast, such as Asbury Harpending, John P. Jones, John W. Mackay, and William C. Ralston, as well as the potent Bank of California. Among the smaller fry in San Francisco it was believed that " 'where Roberts leads, there is the chance for fortune.' "[12]

On this occasion Roberts headed for Leadville with hopes of purchasing and promoting one of the rich properties, then creating a stir in the East. Mindful that his past successes had been partially due to his personally investigating the new mining camps and cultivating the friendship of local mine workers, he chose to board in the bunk house of the Little Pittsburg while he studied the operations on Fryer Hill and elsewhere around Leadville. With Roberts looking on, the Little Pittsburg was then taking out extraordinary quantities of high-grade silver ore, and its neighbors, the Chrysolite and the Little Chief, were steadily progressing and revealing large blocks of ore yet to be mined. Under Tabor's direction, the Chrysolite had already produced over $1 million in profits, and development work had unearthed ore estimated to be worth several million dollars more. Likewise, the Little Chief, after experiencing some trouble with litigation and the high costs of extensive improvements, had started paying

[12] For Roberts' participation in the diamond hoax, see Bruce A. Woodard, *Diamonds in the Salt*, pp. 76–81. *San Francisco Call*, December 25, 1901; Oscar T. Shuck, *Sketches of Leading and Representative Men of San Francisco*, pp. 1001–02; *Pacific Coast Annual*, pp. 48–49; *MSP* 20 (May 14, 1870): 374; *MSP* (September 5, 1874): 148.

dividends to its Chicago owners. Branching out from his makeshift headquarters in Leadville, Roberts also inspected the Freeland mine near Idaho Springs and the OK and Winnebago property at Central City, which, like the Chrysolite and the Little Chief, had been adequately developed in the early stages and were producing silver ore at a profit. Favorably impressed by what he saw, the Californian began negotiations for these four mines and formulated plans for introducing them to the money markets east of Colorado.[13]

While George Roberts busily examined the silver mines of Colorado, the chances of selling a mining property in the East had improved tremendously over 1878. A prime factor in the change was the resumption of specie payments in January, 1879, that bolstered investors' faith in the economy, hurried the return of prosperity, and created a financial climate suitable to the flotation of silver-mining certificates. Less than a year previously, the silver-mining regions of the West had received another and more direct boost from Washington in the Bland-Allison Act, which provided for the partial remonitization of silver. This, coupled with specie payments, made the future of mining investments appear rosy in 1879. A New York stock review added grist to the mill by reminding eastern businessmen of the enormous potential for wealth that mining alone possessed: investments in commercial banking were unattractive; manufacturing, unless monopolized, paid modest returns; merchandising was "precarious"; and farming, though safe, was always a "dull, plodding business." It followed that the chances of an "intelligent capitalist's" becoming a millionaire were ten thousand times better in mining than any other financial pursuit. Sharing the optimism, the *Engineering and Mining Journal* predicted the rapid development of New York into the center of "legitimate" mining investment, leaving to San Francisco the "speculative and most hazardous enterprises."[14]

First of the major Colorado mines to hit the eastern market was the Little Pittsburg Consolidated Mining Company, organized by Chaffee, Moffat, and Tabor and capitalized at a mammoth $20 mil-

---

[13] *Rocky Mountain News*, September 4, 1879, September 12, 1879, December 2, 1881; Hall, *History*, 2: 339–40; *EMJ* 28 (October 18, 1879): 286; *EMJ* 29 (February 7, 1880): 95.

[14] *United States Annual*, p. 17; *EMJ* 27 (February 22, 1879): 136.

lion in 200,000 shares with a par value of $100 each. A triumphal
tour of the East by Tabor, recognized at once by eastern businessmen
as graphic proof of Leadville's wealth, enabled the company to sell
$50,000 of its stock by the spring of 1879 and to be fitted out with
a resplendent board of directors headed by Henry Havemeyer, a future organizer of the sugar trust, A. J. Dam, owner of the swank
Astor House on Fifth Avenue, and former Connecticut senator William H. Barnum. On the strength of high production at the mine and
the opening wedge made in the market by Tabor, Moffat, and Chaffee,
perhaps accompanied by John P. Jones, arrived in New York in May
to sell another $50,000 of stock at $25 a share. To aid in their new
promotional campaign, they retained the services of Charles C. Dodge,
heir to the Phelps-Dodge copper fortune, and Edward H. Potter, a
New York stock broker, both being experienced mining promoters
and, most importantly, having access to local financial circles.[15]

In his capacity as president of the Little Pittsburg, Chaffee, an
old hand at mine financing, chartered a hotel railway car—one which
combined parlor, dining, and sleeping facilities—stocked it with
food, whisky, and cigars, and invited his board of trustees, interested
capitalists, and several journalists to visit his great silver-producer
on Fryer Hill. Deftly planned to create a mood of opulence and
genial optimism among the passengers, the trip aimed at selling more
shares to current holders and to stimulating greater public interest in
the mine that was already leading Colorado's resurgence on the stock
market. Although hundreds of easterners had inspected the Rocky
Mountain mines since 1860, nothing quite matched the group as-
sembled by Chaffee, which left from Jersey City on May 21, 1879,
soon after the company declared its sixth dividend of $100,000. Vice-
president David H. Moffat grandly hosted an entourage that included
West Virginia politician and capitalist Stephen B. Elkins, New York
Stock Exchange president Brayton Ives, prospector George Fryer, Yale
paleontologist O. C. Marsh, speculator Ulysses S. Grant, Jr., engineer

---

[15] The useful reports of correspondent Charles H. Dow to the *Providence Journal* about the promotion of the Little Pittsburg between May 28, 1879, and July 20,
1879, appear in the appendix of George W. Bishop, Jr., *Charles H. Dow and the
Dow Theory*, pp. 250–55. *Denver Tribune*, May 27, 1879; *MSP* 38 (January 4,
1879): 5.

Rossiter Raymond, and newspaper reporters from New York, Boston, and Providence. The skillful blending of present and prospective investors, brokers, and journalists, twenty-eight people in all, from cities that hungered for further news about Leadville not unexpectedly earned the admiration of the *Bullion*, which proclaimed it a "napoleonic move" by the officers of the Little Pittsburg.[16]

Indeed, the plan worked to perfection. Capitivated by the excitement of Leadville, the eastern businessmen promptly purchased 80,000 shares after Raymond and Winfield Scott Keyes, a California mining engineer, issued favorable reports on the mine. In "as fair an appraisal as I can make," Raymond estimated that, apart from the ore dumps, the Little Pittsburg contained $2 million of silver ore in sight, with another $1 million in "probable yield," and, realizing that two-thirds of the property remained unopened, that the "indications of value are favorable, but not yet definitely developed." Keyes dropped some of his professional reserve to exclaim that the New Discovery shaft held the " 'biggest pot of money in the world.' " Present to hear both reports, newsman Charles H. Dow, better remembered for his stock-market theories, did some figuring on his own and reported to his Providence readers that the mine was worth approximately $54 million and that there was enough high-grade ore to pay $100,000 in dividends for three or four years. Extraordinary statements of this kind, even though they distorted Raymond's report, were not lost on eager investors in the East who quickly made the Little Pittsburg a favorite on the stock market and transformed New York into a center of feverish mining activity unknown to the city since the days of 1863–1865. The conservative *New York Times* noticed this "unwholesome" similarity to the earlier Colorado booms and snorted that "there is nothing new in the method, and we apprehend that the result will be equally as stale." It went on to warn easterners that even the finest mining engineers could not "possibly guarantee" the future of a mine, particularly when speculation, overcapitalization, deceptive

---

[16] According to George Upshur, a close friend of Comstock millionaire James C. Flood, U. S. Grant, Jr., was a heel. He reputedly foresaw the decline of the Comstock and the rise of Leadville and, though near to marrying Flood's daughter in 1879, threw his fortunes in with the Little Pittsburg and wed Fanny Chaffee instead (*As I Recall Them*, p. 97). Bishop, Jr. *Dow*, pp. 251–54; *EMJ* 27 (May 24, 1879): 380; *Bullion* 2 (November 1, 1879): 1.

promotion, and mismanagement would almost certainly make the chances of regular dividends "remote."[17]

However, investors paid little attention to such jeremiads as the ballyhoo over mining stocks grew louder in the summer and fall of 1879. The arrival of San Francisco mining operators—men like George Roberts, who had been gathering in the western mines through options and purchase—were well timed to take advantage of this development, and they appeared in New York to sell their schemes or "advise" on the proper nurturing of a mining boom. On both coasts this eastward migration drew mixed reactions. In San Francisco, mining men reluctantly admitted that Leadville had great potential but pointed out that it was too cold, too high, too rainy in summer, and its residents were susceptible to pneumonia and other pulmonary ills; in short, they concluded, with the many opportunities available further West, there was "very little in the Leadville mines that ought to greatly excite the California capitalist, speculator, or prospector." Two months later, as the role of Californians became clearer and speculations ran rampant, a Pacific Coast journal commented philosophically: "Californians having been milked to the strippings, it is meet that the milking machine be taken hence [New York] to perform a little service where the lacteous fluid remains still abundant." New York also expressed ambivalent feelings about its impending role as the mining capital of the nation or its attractiveness for the California operator. Seeing the trend in 1878, Ross Stone feared that a reckless breed of speculators would descend upon his city, although he welcomed with open arms those Californians, mainly mining engineers and experts, who might restore public confidence in the mining industry. A New York mining review saw a rare opportunity to seize control of all the western mines with the aid of these Californians, and it asked the East not to judge San Francisco by a few thieves and charlatans, for to do so would be tantamount to holding Trinity Church accountable for the Black Fridays on Wall Street.[18]

17 Bishop, *Dow*, pp. 301–05; *EMJ* 27 (June 28, 1879): 462–65; *EMJ* 28 (October 25, 1879): 293; *New York Times*, July 5, 1879, p. 4.

18 Not everyone disagreed with the *Times*. The mercantile agencies in Boston informed local businessmen that the purchase of mining stocks would jeopardize their credit ratings (*EMJ* 27 [May 17, 1879]: 361). *MSP* 38 (May 3, 1879): 281;

The hopes expressed in some quarters that this new excitement would learn from the past and that investors would avoid making ruinous mistakes again were as characteristic of a mining-stock boom as the reckless speculation and the blundering into bad risks that invariably occurred. New York, after all, had been through mining booms before—Georgia in 1835, Michigan copper in the 1840s, California in 1850, and Colorado in 1863–1865—when careless investment and ignorance of mining had cost millions of dollars in wasted capital. Reportedly only one company organized in New York during the Colorado rage of the 1860s still survived in 1879 and produced ore and returned dividends. Because of that experience, New Yorkers had become cautious about mining securities, and the better brokers had handled stocks less liable to sudden and dismal failures. In 1879 the *New York Times* thought that the discovery of high-grade silver at Leadville and the return of business prosperity guaranteed another mining fever in the East, but now absentee investors would wisely demand safeguards against worthless, manipulated mines that represented nothing more than poor gambles and big losses. This time New Yorkers sought "mines which pay dividends; which have been developed; which have the speculative element most eliminated from the management; which, in short, offer a reasonable chance to the investor if he desires to make his investment permanent." Moreover, prior experience in mining had created a "class" of intelligent investors, said the president of the New York Mining Stock Exchange, who will "check fraud and imposture, and render its inception dangerous within the range of their criticism."[19]

If, in fact, eastern investors had matured and now demanded proved properties, George Roberts had what they wanted, as, of course, every good salesman must. Aware of the lip service being paid to cautious, sound mineral investments and probably influenced by the market success of the Little Pittsburg as a paying proposition, he had prepared a list of "Roberts dividend-paying mines," supposedly the best of eight hundred inspected in the West, and brought it to New York in the winter of 1879–1880. Between October, 1879, and

---

MSP 39 (July 12, 1879): 24; *United States Annual*, pp. 19–24; Ross C. Stone, *Gold and Silver Mines of America*, pp. 19–23.

[19] *New York Times*, July 8, 1879, pp. 1–2; *Mining Record* 5 (May 24, 1879): 427.

March, 1880, Roberts became the prime mover in organizing the Chrysolite, the Little Chief, and the Iron Silver mining companies, all at Leadville; the Freeland Mining Company at Idaho Springs; the Robinson Consolidated near Breckenridge; and the Winnebago-OK Mining Company at Central City. In each case the mine had been adequately developed and was making a profit—such as the Robinson, with ore that ran $300 in silver to the ton—before its formation into a public joint stock corporation. Likewise, the officers and management of each revealed a number of Roberts' cohorts from California and Nevada, and his mining experts were usually Winfield Keyes and a friend of thirty years, "General" J. B. Low. Conspicuous also were the names of John P. Jones, Henry Rosener, Edward B. Dorsey, and George Daly, all of Comstock fame, and a few eastern allies such as Drake DeKay, a New York commission merchant, and Stephen V. ("Deacon") White, a stock dealer and subsequently congressman from New York. All of the mines carried a large paper capitalization of several million dollars, from $2 million for the Winnebago-OK to $10 million for the Chrysolite, Little Chief, and Iron Silver, and a similar approach was adopted in promoting the stock of each.[20]

The Chrysolite quickly outshone the rest owing to its enormous potential for wealth and its nearness to the flourishing Little Pittsburg. Although Roberts remained discreetly in the background, his stamp appeared on every stage of the promotional campaign, beginning with the appointment of an impressive board of trustees meant to instill public confidence in the enterprise and to push along the sale of stock. Guaranteed to catch investor attention was the name of Horace Tabor, now lieutenant-governor of Colorado, who had traded his interest in the Chrysolite for approximately 70,000 shares of capital stock in the new corporation, making him a trustee along with Ulysses S. Grant, Jr., John P. Jones, publisher Daniel Appleton, and New York banker Henry A. V. Post, the last two being president and vice-president respectively. Consistent with Roberts' strategy, the Chrysolite issued its capital stock fully paid, meaning that the company had purchased the mine with its stock, and became a profit-

[20] *Bullion* 2 (November 15, 1879): 1–3; *Bullion* 2 (March 16, 1880): 1; *Rocky Mountain News*, May 30, 1880; *EMJ* 29 (January 24, 1880); 63; Fossett, *Gold and Silver Mines*, pp. 375–76, 427–29, 451, 496–98.

maker by paying $200,000 in dividends in each of the three months under company control, or $1.00 on each of its 200,000 shares. Furthermore, its prospectus claimed that the company had taken out over 7,000 tons of silver ore with an average net profit of $75.10 per ton.[21]

Yet the wise investor never bought stock in a mine solely on the basis of how much ore had been removed from it. Rather, he would want to know how much silver ore remained and whether there was enough of it to bear the high costs of labor, machinery, and reduction, as well as to show a hefty margin of profit. Ordinarily, the most reliable source of information on these matters appeared in the report of a qualified mining engineer who had personally examined the mine. If one expert was good, then three were better, thought Roberts, who employed Joel Low, Winfield Keyes, and the redoubtable Ross Raymond to visit the mine, inspect the workings, sample the ore body, and generally determine the long-range value of the property as an investment. Accordingly, it became Low's "firm conviction" that the Chrysolite would yield a net profit of $200,000 a month for at least two, but probably four, years. And Keyes, who happened to be general manager of the mine, submitted a flattering comparison of Fryer Hill with the Comstock in which he foresaw no problems in excavating one hundred tons of ore each day from the mine at an "average" gross return of $120 a ton, and he concluded that there was $5 million worth of ore in sight. Finally, Raymond, working for a handsome $5,000 fee, took time away from his inspection of the Little Pittsburg in June, 1879, to examine the Chrysolite, with general manager Keyes providing a running account of the mine's past, present, and future prosperity. That same day Raymond wrote his report on the property and, while acknowledging some dependence upon "those in whom I have confidence," believed the Chrysolite worth $5 million even though that "figure is not based wholly upon personal knowledge." Reprinting the three glowing reports in its prospectus of January, 1880, the company offered its own prognosis: in view of past performance the Chrysolite would yield a product of $74,194,000![22]

Armed with this information, Roberts took the battle to the eastern stock exchanges. To dispel any notion that the mine might be

---

[21] *Chrysolite Silver Mining Company*, pp. 3–7.

[22] Ibid., pp. 7, 17–27. Keyes was also general manager of the Little Chief and the Iron Silver mines.

overpriced for self-enrichment, Roberts devised a slick scheme to sell the capital stock, par valued at $50 each, for $13.89 a share, or exactly the $2,778,000 that the company "paid" for the property, but he reserved an option or "call" to buy back one-half the stock within eight months after refunding the purchase price plus interest. Dark though the purpose was, the *Bullion* received the plan favorably, viewing it, as many investors would, as a sign that the promoters would make a profit only if the mine prospered and neglecting to see that if the enterprise failed Roberts would refuse the option and leave the stockholders with a bundle of depreciated securities. At the time, however, that would have been considered a querulous remark, since the Chrysolite continued to deliver monthly dividends of $200,000 and to quicken its tempo of stock sales. By the end of January, its stock had soared from $13.89 to $40 a share, thus nearly tripling in value over a period of four months, and part of the credit for this went to Keyes in Colorado, who faithfully fed a stream of ore receipts and good news into the New York office, where Roberts' genius for publicity took over. According to one report, Keyes excavated 1,130 tons of ore during one week in January that paid nearly $100,000 into the company treasury. Out of such stuff Chrysolite shareholders built dreams of becoming silver kings.[23]

With the Chrysolite and the Little Pittsburg setting the pace, the East rambled into its greatest mining boom of the century, unmatched for its intensity, the vast sums of money involved, and the perfect blending of western bonanzas with eastern prosperity. Even the editor of the *Mining and Scientific Press*, accustomed as he was to Comstock splurges, rose to new heights of eloquence over the size of this boom:

The miner's pick is more potent now than was the magician's noted talisman of bygone ages. At its practical stroke, from the darkness of the mine it brings to light the yellow gold, the glistening silver, the bright brown copper, the more ductile lead, and the somber-hued iron. It opens receptacles of hidden treasures, that since creation's dawn have awaited its steady blows to break away the casement that enclosed them. It wields more power than did the subtle spells of the alchemist eons of years ago, and transmutes the crumbling rock into bright, pure gold. It

23 *Bullion* 2 (November 15, 1879): 2; Ingham, *Digging Gold*, p. 378; *EMJ* 28 (October 11, 1879): 7; *EMJ* 29 (January 24, 1880): 62; *Rocky Mountain News*, December 30, 1879.

sends the unsightly ore to the furnace, from whose fusion is extracted the noble metal.

More to the point, a New York paper estimated that easterners had spent $20 million on the western mines during 1879 and were expected to surpass that figure in 1880. Larger, wealthier investors gobbled up mining securities in huge blocks—some in one-thousand-share lots—while smaller speculators, many of whom were inveterate gamblers, gathered at the stock exchanges each morning to buy a few shares in one of the major companies or to speculate in the latest "wildcat" mine upending the market. Elsewhere, reports indicated that many "members of Congress are plunging into mining speculations," and these were not the old hands like Chaffee and Jones, but a new breed which had never before invested in the industry. Whether politicians or bankers, those in New York could always hear the latest gossip on the boom and haggle over details on a "sure thing" at the fancy Fifth Avenue Hotel, which became the headquarters for promoters, experts, and financiers, who jammed its French renaissance halls each day from 7:30 A.M. to 10:00 P.M. A competing hotel tried to draw the mining crowd to its rooms by redoing the lobby in a western motif, while another, unimpressed by the braggadocio of its mining clientele, erected a hand-lettered sign above the main desk: "No mines taken in payment for board!"[24]

The multitude of stock buyers in 1879 and 1880 caused a swift expansion in the number of exchanges and brokerage houses to meet the demand of "all classes" for mining investments.

Neither age, sex or occupation place any bar upon the action, in this line, of individuals. From the bootblack to the banker, through all the intermediate grades of capital; from the corn doctor to the surgeon of consummate skill; from the constructor of wheelbarrows to the builder of the most intricate steam engines; from the scullion in the kitchen to the millionaire owner of the palatial residence; and through every phase of social life, from the lowest to the highest—all are swelling the grand army of stock buyers, and becoming thereby co-partners in mining properties.

In New York, where the fever became an epidemic, many small, pri-

[24] *New York Tribune* in *MSP* 40 (March 6, 1880): 146. See also *MSP* 38

vate exchanges immediately responded by displaying huge placards announcing that they now carried a full array of mining securities at the latest quotations. Not to let speculation become still another male prerogative, a Mrs. Marion E. Warren organized the Ladies' Mining and Stock Exchange in Union Square and invited chambermaids and dowagers to enjoy the freedom of gambling on the mines. But the excitement most thoroughly altered the somnolent New York Mining Stock Exchange, which had been formed during the Civil War, thrived for a brief time, disappeared except as a legal entity, then revived in 1877 with aid of the powerful New York Stock Exchange. In 1879, it rocketed into the limelight; the cost of a seat on the exchange jumped from $75 to $1,100 and stock sales rose from a few hundred shares a day to 15,851,878 for the year, or an average of 50,000 for every business day. Nor was New York the only city hypnotized by mining. Philadelphia opened two new mining exchanges in less than a year; Boston added a separate mining board in 1880; St. Louis established an exchange in the summer of 1880; and Chicago brokers, at first using large chalk boards to quote their stocks, organized the Chicago Mining Exchange in 1882, with more than two-thirds of its listed companies situated in Colorado. While Leadville mines were also the focus of attention on these exchanges, the companies differed from those entrancing New Yorkers. Philadelphians, for instance, held the "watered" stock of the Iowa Gulch Mining Company in high esteem, and St. Louis investors traced the progress of the Small Hopes and the Colonel Sellers mines. No matter what the local favorite, however, the mining exchanges prospered under the boom conditions and eagerly fed the public's appetite for mining shares.[25]

Though millionaires might be "as thick as raspberries" in New

(April 5, 1879): 216, *MSP* 40 (February 28, 1880 and March 13, 1880): 133, 161; *MSP* 41 (October 2, 1880): 217. *Bullion* 5 (May 9, 1881): 175; *EMJ* 29 (January 12, 1880): 41–42; *Mining Record* 5 (July 12, 1879): 25–26.

[25] New mining areas in Arizona, Utah, and Montana, as well as older ones in Nevada and California, also attracted eastern capital. However, the majority of mining companies formed in New York between 1878 and 1880 were to work in Colorado, according to an examination of the *New York Mining Directory* for 1880. *MSP* 40 (March 13, 1880): 161; William R. Balch, comp., *The Mines, Miners and Mining Interests of the United States in 1882*, pp. 475–76, 496–506; *New York Times*, July 8, 1879, pp. 1–2, and November 13, 1880, p. 2; *EMJ* 28 (December 13, 1879): 44; *EMJ* 31 (February 5, 1881): 98.

York, George Roberts saw a need to enlarge the city's buying power and at the same time avoid some of the opposition to his methods from the New York Mining Stock Exchange. To do this, he proposed the creation of a new mining-stock exchange, along with a banking house, which would encourage speculation by loaning money for the purchase of securities in the same way that San Francisco banks had made money available to Comstock operators. Accordingly, in early 1880, with John P. Jones, Jerome B. Chaffee, Horace Tabor, ex-Arkansas senator Alexander McDonald, former governor of Arizona Territory Anson P. K. Safford, and several others, Roberts began to organize an exchange, bank, and clearinghouse to operate as a unit for the freer buying and selling of stocks, with control of the machinery firmly resting in his hands. Daily meetings were held in the Boreel Building (called the "Bore-al" by some cynics), which contained the offices of the Chrysolite, Little Pittsburg, and many other mining companies, and where Roberts brought together his friends and "household gods" from the San Francisco Stock and Exchange Board, the obvious model for this New York scheme.[26]

Roberts' efforts resulted in the formation of the American Mining Stock Exchange in the winter of 1880 under an open-ended charter that had originally been issued to the notorious Tweed Ring in 1868 as part of an attempt to "blackmail" the New York Stock Exchange during a railroad-stock war. Except for a few specially chosen brokers who paid $1,000 for a seat on the new exchange, membership cost $5,000 and was to be limited to eight hundred. The real success of the whole scheme rested on the Mutual Trust Company, which obtained its loan fund of more than $200,000 from the sale of seats on the exchange and from deposits made by the very same mining companies whose shares the exchange offered for sale. In short, the trust enabled the public to buy approved mining stocks on margins as high as 40 and 50 percent and accepted those securities as collateral on the loans. By providing financial aid for the purchase of their own wares, Roberts and company closed a circle of control over the mines and small investors that encompassed promotion, financing, and management, and extended their opportunities for

---

[26] *New York Times*, February 21, 1880, p. 3, and March 25, 1880, p. 2; *Mining Record* 5 (June 21, 1879): 525; *EMJ* 28 (November 1, 1879): 311.

profit from ore yields to finance charges to the sale of seats on the exchange. And to make certain that the system worked with the efficiency of San Francisco's, California experts dominated the bank and the exchange: Milton S. Latham, ex-California senator and a San Francisco banker, headed the trust; C. T. Christenson of the Nevada Bank of San Francisco became treasurer; George W. Smiley, former chairman of the San Francisco Stock and Exchange Board served as chairman of the new exchange; and Roberts and John Jones were trustees in both institutions. "A gang of wolves" from San Francisco, said one critic, who denounced the bank as a "flytrap, pure and simple, except that the fly is a lambkin and the big black spider is a wolf."[27]

With great fanfare the exchange opened its doors on June 1, 1880, directly across the street from its chief rival in the sale of stock, the New York Mining Stock Exchange. Actually the American Exchange made a grand addition to the financial center of lower Broadway, since all of its rooms had been designed and decorated to imitate the sumptuous San Francisco Stock and Exchange Board, complete with expensive cherry and mahogany wood interiors and gilded railings and spittoons. Once the windy welcoming speeches had ended, the members launched the enterprise with a few hours of brisk trading and bogus sales in the thirty-two stocks (mostly Colorado) that appeared on the board. Appropriately, this first day of business concluded with dinner at Delmonico's, the fashionable restaurant of New York's business community, and, during libations from an "unlimited" supply of champagne, the officers read telegrams from the Comstock "bonanza kings" saluting the new exchange and wishing the organization well in bringing the San Francisco system to the prosperous East. In the opinion of the *New York Times*, which had taken a captious view of the boom from the start, the East would never see the full riches of mining until "the chicken that finds a particularly nice worm" decides to share it with its underprivileged mates. And that time had simply not come.[28]

[27] *Mining Record* 7 (January 10, 1880, and February 21, 1880): 25–26, 174; Frederick H. Smith, *Rocks, Minerals and Stocks*, p. 211; *New York Mining Directory*, pp. 110–11; *The Mining Trust Company and Its Exchange Building*, pp. 3–5, 31–33; *New York Times*, February 21, 1880, p. 3, and February 27, 1880, p. 8.

[28] *New York Times*, June 2, 1880, p. 8, and June 19, 1880, p. 4; *Mining Record* 7 (June 5, 1880): 549; *Bullion* 3 (July 12, 1880): 12.

Even as Roberts pushed forward his plans to nourish the excitement, the rapid and mysterious decline of the Little Pittsburg jolted the market. Prior to January, 1880, the mine had paid a total of $850,000 in dividends and its stock sold well at more than $30 a share, when suddenly, and despite continued ore production and dividends, its stock plummeted to $15 by the end of February and to a mere $5 by mid-May. The first diagnoses of the trouble centered on a bear movement or a deliberate attempt by company officials to devalue the stock for speculative reasons, and the "past history of some of the insiders would naturally expect one to look for an occasional trick." In fact most of the reportedly 85,000 shares of Little Pittsburg sold in February and March had been held by David Moffat and Jerome Chaffee, who secretly deserted the enterprise. Years later, Chaffee confessed to selling his stocks early in the decline, but he argued that he had been hurt as badly as anyone else and added, somewhat defensively, that his own daughter (Mrs. U. S. Grant, Jr.) had lost eleven dollars on every share she owned.[29]

Once proclaimed to be "as strong as the Bank of England," the Little Pittsburg now caused bitter resentment among easterners who tried to locate the reasons and persons behind this "unwholesome and highly respectable robbery." The hunt for a scapegoat eventually focused on Raymond, the examining engineer, whose favorable report on the property had been used, sometimes deceptively, to promote the stock. Amid a crescendo of accusations that he had overvalued the mine, shielded the public from its true condition, and even speculated in the stock himself, Raymond released another report on the property to explain the sudden drop in ore production. After a visit to Leadville, the engineer concluded that development work on the Little Pittsburg had been "crippled and almost prevented" by the priority given to removing the highest grade ore to pay large and speedy returns on the capital investment, and that, in his view, the management had sacrificed careful, common-sense mining to the demands of the stock market. He earnestly dismissed as a "mere vulgar falsehood, and not funny at all" the charge that he had originally reported $5

---

[29] Tabor reportedly sold his interest in the Little Pittsburg to Moffat and Chaffee for approximately $1 million before the collapse came (Gandy, *Tabors,* p. 193). *EMJ* 29 (February 28, 1880): 156; *Bullion* 2 (March 1, 1880): 5; *New York Times,* April 23, 1885, p. 8.

million of silver ore in sight, and he denied that he or his friends had profited in any way from the wild speculation.[30]

From the available evidence it seems that the Little Pittsburg suffered a fate common to the precious-metal mines of the West— vigorous exploitation of its high-grade ore deposits at the expense of slow, steady development. Having driven up the price of company stock by producing bonanza amounts of silver and declaring big dividends, the management obviously found it difficult to slow down, reduce returns, and put more money into long-range projects at the mine, without causing a volatile reaction on the eastern market. As the time neared when the property could no longer sustain such a system, Moffat and Chaffee, who had a combined experience of about fifty years in Colorado mining and were always privy to the weekly reports of superintendent J. B. Bearce, began to pull out. One journalist had learned from Bearce as early as September, 1879, that the pressure for rapid production was ruining the Little Pittsburg and that the reserves of rich ore were being removed faster than he could open, explore, properly timber, and develop new bodies of ore to maintain monthly dividends of $100,000. However, with these facts available to insiders, the company stressed in early 1880 that only 5 percent of the total property had been opened and claimed that $5 million worth of silver ore was in sight and that an additional $100,000 in ore would be mined each month.[31]

In a suit that took several years for a verdict, a New York shareholder brought Moffat and Chaffee to court for his losses in the Little Pittsburg, alleging that the Coloradans had misrepresented the mine for their own profit. During the trial in the New York State Supreme Court, the pair defended their honorable conduct against charges of deceit; Moffat asserted that he personally inspected the mine and found it enormously valuable, and Chaffee, admitting under cross examination that the prospectus may have been "injudicious—too extravagant," still thought the Little Pittsburg could pay $100 million. After deliberating all night April 30, 1885, the jury returned a verdict of not guilty for "wilful misrepresentation," and Moffat and

[30] *New York Times*, July 21, 1879, p. 5; *EMJ* 29 (April 3, 1880): 232–34, 236; *Bullion* 2 (March 1, 1880): 5; Ingham, *Digging Gold*, pp. 389–90.
[31] *EMJ* 29 (March 6, 1880 and March 20, 1880): 76, 199, 201; Ingham, *Digging Gold*, pp. 379–80; Hall, *History*, 2: 438, 455–58.

Chaffee, smiling "jubilantly," returned to Colorado. They had good reason to be happy. Early in the trial they had brought together many of the miners who had worked in the Little Pittsburg in 1879 and 1880, treated them to dinner, whisky, and cigars at a total cost of $195 and, having them in an amiable mood, asked them to recall for a stenographer and a lawyer their reminiscences of how promising the mine looked shortly before its sudden collapse. The man who handled this bit of skulduggery in Leadville, Oliver H. Harker, informed Moffat that he treated the witnesses "very cleverly to get what was wanted," and added that he could not "see how you can lose the suit with the testimony furnished." Moffat and Chaffee did establish their "legal" innocence, but there is little doubt that their management and stock operations made them culpable for the collapse of the Little Pittsburg in 1880 and the disruption of the stock market that followed.[32]

If any mine could replace the Little Pittsburg in the hearts and purses of the eastern public, it was the Chrysolite, and from every indication in the winter of 1880, this seemed to be its destiny. Shares sold at $40 each, and George Roberts devoted his every effort to advancing the stock by making lusty predictions of high returns for the future. Early in April, the company declared its monthly dividend of $200,000, making a grand total of $1 million paid to stockholders since its organization six months earlier. Besides having $7 million to $8 million of silver ore in sight according to company reports, the mine, under the skillful supervision of Winfield Keyes, had become "incontestably the best-managed mine in the district." Thus, it appeared that the Chrysolite might still fulfill many of those spectacular promises made in the East about the carbonate fields of Leadville.[33]

Then, in mid-April, with the market demoralized by the Little Pittsburg debacle, Chrysolite stock abruptly fell to a lowly $15 a share. On top of this shock came word from the company that dividends would be reduced by one-half in order to keep a cash surplus

[32] Harker managed Moffat's Maid of Erin mine at Leadville. Harker to Moffat, January 12, 1885, David H. Moffat, Jr., Papers, First National Bank of Denver Archives. *New York Times*, October 18, 1884, p. 271; April 23, 1885, p. 8; April 29, 1885, p. 8; and May 1, 1885, p. 2.

[33] *EMJ* 29 (March 13, 1880, and March 27, 1880): 189, 220; *Rocky Mountain News*, March 13, 1880; Fossett, *Gold and Silver Mines*, p. 447.

on hand because of poor road conditions in Leadville that delayed ore shipments and because of the recent purchase of the Vulture claim from Horace Tabor, himself a stockholder, for the tidy sum of $250,000. Granting that these actions would depress any stock, one New York paper argued that the stock had been "beared" by the large shareholders who had dumped their stock on the market, depressed it further by cutting dividends, and intended to repurchase later at a lower price. According to this view, manipulation and panic selling explained the Chrysolite's troubles. Keyes seemed to concur when he labeled as "palpable lies" reports that the mine had been exhausted, and he further pledged to maintain production.[34]

Instead matters became considerably worse. On the morning of May 26, the Chrysolite miners called a strike, walked off the job, and marched from mine to mine inviting their fellow workers to join in a protest against labor practices of the Leadville mining companies. Ostensibly, the conflict, which began at the Chrysolite, involved typical employee grievances—higher wages, shorter working days, and the dismissal of several miners by the company—but in the background loomed the miners' anger at the rigorous enforcement of new regulations against smoking and talking underground. To these complaints must be added the general uneasiness among all workers in Leadville that its future as a productive camp might be no more certain than that of the sluggish Little Pittsburg. In the face of these problems, George Daly, a long-time associate of Roberts and the assistant manager of the Chrysolite, adamantly refused to negotiate and, instead, assumed an heroic stance among the mine owners by firing the strikers and replacing them with new men at the old wage rate of three dollars a day. Furthermore, he and Keyes, acting on orders from New York, reinforced the Chrysolite shaft house with iron pillars, turning the mine into a fortress, and posted armed guards above and below ground to ward off threatened attempts to sabotage the company's pumps and flood the workings. While most other properties shut down to wait out the strike, the Chrysolite took a defiant position by keeping 450 men at work and producing ore, although doing so at tremendous expense and under the greatest dis-

---

[34] Article from the *New York Commercial Advertizer* appeared in *Rocky Mountain News*, May 7, 1880. *EMJ* 29 (April 17, 1880, April 24, 1880, and June 5, 1880): 276, 290, 292, 383–84.

advantages. After repeated demands from the mine owners—Daly and Keyes foremost among them—Governor Frederick Pitkin declared martial law in mid-June and placed Leadville under the control of the Colorado militia, which soon broke the back of the strike.[35]

To some informed observers it was no mere coincidence that the strike began at the Chrysolite or that its management took a belligerent stand against the strikers. Carlyle C. Davis, editor of the *Leadville Democrat*, recalled that it was "alleged and generally believed" that Daly and Keyes had sown the "seeds of discord and discontent" among the miners and brought on a strike to serve the interests of stock manipulators in the East. He also pointed out that local feelings ran so strongly against the pair for what they had done that "continued residence in Leadville could not be considered with pleasurable anticipation," and this probably explains their hasty departure from the town soon after the militia left. During the strike, Daly admitted to the *Rocky Mountain News* that the trouble indeed originated in the Chrysolite and, though never mentioning any role for himself, stated that the shift bosses had received strict orders from the company to enforce a variety of new rules and regulations unpopular with the miners. At a later date, the influential *News* laid much of the blame for the deployment of militia in Leadville upon the Chrysolite managers and strongly implied that the state's power and machinery had been used for unsavory purposes, adding that, "We made those men. They have unmade us, with the strength we gave them as a credulous community. . . . " Events which followed heightened suspicions that the company, acting mainly through the hard-nosed Daly, had precipitated a strike, inflamed passions by rejecting negotiations and calling for martial law, and caused a great disruption of the Leadville mines, so that operators on the eastern exchanges could drive stock prices down.[36]

[35] Hall, *History*, 2: 460–64; Richard E. Lingenfelter, *The Hardrock Miners: A History of the Mining Labor Movement in the American West, 1863–1893*, pp. 143–56; *EMJ* 29 (June 5, 1880, and June 12, 1880): 383–84, 410; *Rocky Mountain News*, May 29, 1880, and June 6, 1880; Vernon H. Jensen, *Heritage of Conflict: Labor Relations in the Non-ferrous Metal Industry up to 1930*, pp. 19–24; Charles Merrill Hough, "Leadville, Colorado, 1878 to 1898: A Study in Unionism" (Master's thesis, University of Colorado, 1958), pp. 46–57.

[36] Daly, an implacable foe of the miners' unions, was described by an acquaintance as a "creeping, crawling, bung-sucking sycophant," who knew "ab-

Shortly before the strike began, Roberts had announced that the Chrysolite would halt regular monthly dividends in favor of quarterly ones, because the old system drained cash reserves and because it caused panic selling by small stockholders when their dividends failed to arrive precisely on time. He stunned the East further by reporting his company in debt for $80,000. Coupled with the news reaching New York about the labor strife in Leadville, the price of stock, which had been holding rather firm on the American Exchange, started to tumble rapidly and hit a low of $6 a share in the summer, as still more rumors circulated that several of the trustees had secretly resigned and left on vacation. The first explanation of the company's troubles emphasized the large sums of money borrowed to defend itself against union saboteurs and to proceed with ore production under the most expensive and hazardous circumstances. But to answer fully the persistent questions about conditions at the mine and its future, Roberts journeyed to Leadville at the end of July to give his personal and expert opinion. After touring the property in a "flippant manner," said a local newspaper, he brusquely informed New Yorkers that the Chrysolite was "finished" as a profitable enterprise, its ore bodies depleted and the company saddled with a mammoth debt now nearing $400,000. Like its neighbor on Fryer Hill, the Little Pittsburg, the Chrysolite had simply run out of pay ore.[37]

Not many in the East, and especially not those who closely followed the progress of the mines, agreed with this simple and self-serving analysis given by Roberts. The loudest objections came from the prestigious *Engineering and Mining Journal*, which felt that the Chrysolite management deserved the "severest criticism" for having used a potentially great mine for its own selfish and dishonorable purposes on the stock market, and it angrily accused the "Roberts combination" and its "California system of management" of single-mindedly looting the mine of its high-grade ore and skillfully manip-

---

solutely nothing about practical mining, and less, if possible, of the proprieties and decencies of good citizenship" (quoted by Lingenfelter, *Hardrock Miners*, p. 137). Carlyle C. Davis, *Olden Times in Colorado*, pp. 249, 261; *Rocky Mountain News*, May 27, 1880, and August 8, 1880; *Mining Record* 7 (June 12, 1880): 573–74.

[37] *Leadville Chronicle* article on Roberts' visit in *Rocky Mountain News*, July 30, 1880. *Mining Record* 7 (May 15, 1880): 460; *EMJ* 29 (May 15, 1880): 343; *EMJ* 30 (July 17, 1880 and July 31, 1880): 37, 77.

ulating the stock to benefit a few at the expense of the many. Rhetorically, the journal asked, "which did the insiders make the most at —booming the stock up to $40 or bearing it down to $6?" Unfortunately, the question could never be answered, although most critics needed no reply from Roberts to feel certain that he and his cohorts had gobbled up the lion's share of the Chrysolite's wealth, whether on Fryer Hill or Wall Street.[38]

As could be expected, the hail of criticism cut a large swath. The motives of men of good will were impugned and their character and integrity publicly assailed because of their association with the malodorous enterprise. Once again the name of engineer Raymond appeared on the list for censure, since he had not only given the Chrysolite a high recommendation in the summer of 1879, but had returned to the mine in March, 1880, and reported the property in good shape, with approximately $7 million in silver ore to be removed. In private, Raymond seemed no less sure that the mine had a great future, and he approved purchase of the stock by his close friends and business associates, Peter Cooper and Abram Hewitt, who invested heavily at $30 a share. Thus, when the Chrysolite suddenly disintegrated, Raymond became a prime target of irate journalists and stockholders who believed he had abused his trust by misleading the public into buying overpriced shares and, worse yet, convinced it to keep those securities as the company neared collapse. He was accused of not knowing the difference "betwixt ore and sandstone," and Ross Stone of the *Bullion* chided those investors who followed the engineer's advice after the fiasco of Little Pittsburg. To add to Raymond's embarrassment Sidney DeKay (the brother of Drake, secretary of the Chrysolite) sued him in April, 1881, for $21,000, which he allegedly lost on 1,700 shares of stock purchased on the strength of the mine examination. DeKay charged that the examination had been careless and that the mine actually contained $500,000, not the $7 million estimated by Raymond. Although the New York court did not hold him liable for pecuniary losses on the stock, the engineer and the profession that he represented more often and better than anyone else suffered a deep wound that would take time to heal. Of course, any professional mining engineer who publicly released his expert opinion on a mine that

[38] *EMJ* 30 (July 24, 1880, and August 14, 1880): 61, 105.

became the object of intense speculation necessarily exposed himself to condemnation when the property failed. He also became the convenient scapegoat for incompetent or dishonest operators on one side and the target of aggrieved stockholders on the other. In Raymond's case, he did overvalue the Chrysolite, but, more seriously, he apparently used company-supplied information as he hurriedly completed his examination. On both occasions, he accepted ore samples from the mine management and did not thoroughly test a huge block of ore, which eventually proved to have a core of worthless limestone. The poor information he received and his own unprofessional and irresponsible conduct resulted in an erroneous, though not a "deliberately" false, report.[39]

Raymond's faith in the long-term profitability of the Chrysolite remained firm, and in October, 1880, he accepted the presidency of the company from a new board of directors stubbornly trying to fend off stock-market bears. Unexpectedly, the "stock jobbers," who were variously associated with the old management, got some help from a ruinous fire in the underground workings of the mine. The fire spread rapidly and burned for more than a month, leaving the stopes and tunnels filled with carbon dioxide and black, acrid smoke. At first, rumors connected the fire with speculative maneuvers on the eastern exchanges, but later evidence indicated that it began accidently when a miner's candle overturned and ignited the dry and irregularly set timber shafting, and only in the sense that the mine had been poorly developed could the past management be blamed for this disaster. The *Engineering and Mining Journal* hurriedly moved to clear away any doubts about the new president. "Formerly," it said, "the mine was all right, and it was the management that suffered from gas. Now, the management has been ventilated and purified and the gas has taken to the mine." In fact, under Raymond's direction, which lasted until 1899, the Chrysolite paid off its debts and returned good profits, thereby vindicating his belief that a good mine had been recklessly managed for the profit of a few.[40]

[39] Raymond was the consulting engineer for Cooper and Hewitt, owners of one of the largest iron and steel mills in the country. Henry B. Clifford, *Years of Dishonor or the Cause of the Depression in Mining Stocks, also the Remedy*, p. 10; *MSP* 42 (April 9, 1881): 232; Thomas A. Rickard, ed., *Rossiter Worthington Raymond*, p. 56.

[40] In November, 1880, the board of directors included Raymond, Abram

Speculative management and stock-market manipulation seem to have been the chief contributions made by George Roberts and many smaller operators like him to the great mining boom of 1879 and 1880. While the Chrysolite was the most spectacular example, stockholders and others voiced similar suspicions about his conduct in the Iron Silver, Little Chief, and Robinson Consolidated mining companies, all of which fell apart in 1880 and 1881. The Iron Silver, which had produced nearly $500,000 under its original owners, paid no profits under Roberts, who instead spent a $170,000 cash surplus and created a debt of $30,000 before bowing out of the enterprise. At the time he announced his departure from the Little Chief, George Daly reported that the mine had been gutted, and Roberts, who had been vice-president, gingerly surrendered the property to its bewildered figurehead president, General Adelbert Ames. The Robinson Consolidated exhibited the same classic symptoms of mismanagement and trickery—its rich ore was removed and the mine left in debt, part of the latter due to Daly's personal loans, which added $31,000 to company expenses. "Lots of fellows got beautifully stuck in the damned thing," said an embittered Robinson stockholder.[41]

As for the Roberts-dominated American Mining Stock Exchange and Mutual Trust Company, they followed a similarly besmirched path to their end, along with his "dividend-paying" mines. Leaving behind a legacy of denunciation that it force-fed bad stock to the public and called in its loans when investors stood to lose the most, the trust company silently vanished from New York around 1883. Brokers on the exchange also learned that their membership fees, which were supposed to have been converted into United States gov-

Hewitt, Daniel S. Appleton, New York politician Thomas C. Platt, and Charles Francis Adams, Jr. Charles M. Rolker was mine manager at the time of the fire; see his "Notes on a Fire Bulkhead," *Transactions of the American Institute of Mining Engineers* 13 (February, 1884–June, 1885): 505–07. *EMJ* 30 (October 9, 1880, November 20, 1880, and November 27, 1880): 233, 329–30, 346–47; Davis, *Olden Times in Colorado*, p. 200.

    [41] *Mining Record* 8 (December 4, 1880): 536–37; *EMJ* 30 (August 14, 1880, and September 11, 1880): 113, 177; *EMJ* 31 (June 25, 1881): 429–30; *EMJ* 35 (June 2, 1883): 312; *MSP* 44 (January 21, 1882, and April 29, 1882): 34, 277; William H. Bush to Horace Tabor, May 22, 1882, Folder 9, Horace A. W. Tabor Papers, Colorado State Historical Society, Denver.

ernment bonds and held in the name of the members, had been spent and lost, probably in speculation, by the trust company. The entire scheme proved to be one grand swindle that victimized the public through watered stock and bamboozled small brokers who thought they could share in the spoils.[42]

By 1884, with the bonanza ores of Leadville nearly gone, eastern capitalists retrenching, and the wild enthusiasms of the early days deadened by fraud and failure, the combustible elements in the mining mania had disappeared. In that year the number of mining shares sold dropped by over 9 million. Such an eager and energetic mining booster as Ross C. Stone had seen the depression coming in 1882 and reduced his *Bullion* to a monthly and shifted his loyalties to the railroads and a pet project, a water velocipede, known as the Fryer Buoyant Propellar Steamship. Thomas Ewing, Jr., an active promoter in the Little Chief and the Robinson Consolidated, began to push shares in a telephone company, while many of his cohorts in New York turned to the electricity boom. Of course, it was not in the nature of the promoting crowd to offer postmortems on the speculative bubble, least of all when there were new dreams to build, and so it fell to others to analyze what had happened.[43]

New York mining expert Henry B. Clifford saw the excitement as "years of dishonor" during which a "broken down Nevada manipulator" had scattered the "seeds of dishonesty among those of a speculative frame of mind." Though he never mentioned Roberts by name, Clifford condemned those mining companies Roberts organized for deceiving the public and for being overcapitalized, overpromoted, overworked, and eventually overtaken by inefficiency, deliberate mishandling, and stock laundering on the eastern market. Like Little Eva in *Uncle Tom's Cabin*, who was "too good and virtuous to live in this sinful world," the huge dividends declared by the mines lasted long enough to bull the price of stocks and ended as soon as the promoters

[42] The American combined with the New York Stock Exchange in 1883. *EMJ* 35 (February 24, 1883, and March 24, 1883): 107, 169; *New York Times*, May 29, 1884, p. 8; *Bullion* 3 (August 23, 1880): 59–60.

[43] *EMJ* 39 (January 3, 1885): 12; Balch, comp., *The Mines . . . in 1882*, p. 492; Charles C. Copeland to Mrs. Cyrus McCormick, May 25, 1882, Series 2A, Box 34, Cyrus H. McCormick Papers, State Historical Society of Wisconsin, Madison.

had milked the investing public dry. For Clifford, the invasion of Californians bearing Colorado mines had dealt the single most destructive blow to a mining boom that might have enriched the nation had it not been for the "leprosy" of stock manipulation. After studying the eastern situation in 1881, a California visitor noted the absence of that "gaming spirit" he had known in San Francisco, where dividends meant nothing and profits naturally came from clever operations on the stock boards. New Yorkers he found to be naïve and even "irrational," since they really expected dividends, wrapped mining companies in "intense respectability," and actually believed that stock exchanges served serious investment purposes. Thus, when the boom collapsed, eastern faith in mining was totally shattered, and the stock exchanges became, not merely disreputable gambling dens, but "dingy monuments of broken pledges and disappointed hopes, founded on what purported to be an honest, businesslike system." Finally, a wisely simple appraisal came from a Leadville miner: "In this camp a prospect may turn a mine any day; a mine a bonanza; a bonanza can be mismanaged."[44]

On the positive side, the great influx of eastern capital developed an exceptionally rich mining center at Leadville, one which carried Colorado into the forefront of mineral-producing states in the nation. According to one source, between 1879 and 1883 Leadville produced over $70 million in bullion, or seven times what the area had yielded in the preceding ten years. Because the carbonate beds were an unfamiliar geological formation to miners and financiers accustomed to dealing with fissure veins, and because of the difficulties and expense in reaching the camp, this accomplishment deserves special attention. Furthermore, the excitement opened hundreds of shafts at Leadville, many of which developed into long and consistent producers, employed thousands of men in and around the mines, and literally constructed an industrial city on the raw frontier. And, more importantly, it encouraged capitalists to explore the mineral resources and business opportunities elsewhere in the West and advanced the economic integration of the country. Hopefully, said the *Boston Economist*, even the many mistakes made by the East in trusting un-

[44] Clifford, *Years of Dishonor*, pp. 4–17; *Bullion* 5 (April 18, 1881): 145; Paul Ward to *MSP* (43 [September 10, 1881]: 170); D. Bauman, *King Carbonate, Leadville, Colorado*, p. 14.

scrupulous operators would "prove finally of inestimable value," and
future investments would be more carefully and wisely made.[45]

[45] *Boston Economist* in *MSP* 42 (March 5, 1881): 146; Ingham, *Digging
Gold*, p. 125; George B. Dresher, *A Description of Colorado, Leadville, and the
Sovereign Consolidated Silver Mines*, pp. 133–35.

5

"Caveat Emptor": The Hazards of Mining Investments

"I T was like hunting and war, precarious and full of peril," remarked a government official in 1881 after surveying the unsteady precious-metal mining business in the American West. More exactly, throughout the late nineteenth century, the mining investor faced a maelstrom of financial, legal and managerial problems that seriously threatened the life of his enterprise and diminished the returns on his capital expenditure. At times overlapping and interweaving enough to bar sharp distinction, these persistent adversities, both within and beyond the control of individual investors, imperiled sound and rational business operations and became so common that they typified the industry as a whole. It seems that the more conservative capitalist, understanding the gravity of these conditions, resisted the mines until the problems could be removed or at least minimized, but others, no doubt fascinated by the dream of immeasurable wealth, plunged ahead, hoping to realize their dreams in spite of the stumbling blocks. After all, mused an anonymous investor, mining would always be "pure speculation, with an added gamble."[1]

Along with the industry in neighboring states and territories, the mining business in Colorado shared an ill-smelling reputation for harboring thieves, charlatans, "process-maniacs," and other scoundrels who masqueraded before the public and cheated investors out of their

---

[1] Robert Porter, statistician with the Census Bureau, quoted by George B. Dresher, *A Description of Colorado, Leadville, and the Sovereign Consolidated Silver Mines*, p. 143. "My Mining Investments," *Lippincott's Magazine* 27 (January, 1881): 86.

hard-earned savings. But bear in mind that nearly every case of lost money in the mines opened a Pandora's box of noisy charges alleging trickery and maleficence on the part of promoters and company officers, who, in turn, earnestly proclaimed their innocence. As a result the historian's view is often clouded, and he is left with relatively little in the way of hard evidence to condemn individuals or companies for malicious exaggerations or willful misappropriations of funds as many angry investors charged. Add to this that mining operations were conducted underground out of sight, required engineering and scientific knowledge not personally possessed by investors, and relied upon the liberality of nature, and it can be better understood how a cloak of mystery invariably surrounded the business and made legerdemain easy for dishonest operators, although equally hard to prove by those who lost money.

Thousands of investors who suffered pecuniary losses in Colorado could have traced part of their misfortune to the flimsy financial structures erected by mine-company organizers. In an age when Americans saw more than a few "shoestring" operations in railroading, town building, and elsewhere, western mining companies earned a dubious distinction for defaulting on their investors and generally having a high rate of attrition. In fact, hundreds of firms organized to work in Colorado never advanced beyond the point of acquiring a corporate charter, printing a prospectus and stock certificates, and selling a few shares to the public, or they disintegrated in the early stages of driving a deep tunnel or constructing a mill. What money may have been invested in these still-born operations usually disappeared into the pockets of promoters, leaving shareholders angry at the business of mining and at themselves for not examining the viability of the scheme more carefully. Norvin Green unhappily acknowledged in 1892 that he had loaned one such enterprise $6,000 and bought "more stock than anyone else," only to lose his entire investment when the project failed to materialize.[2]

Still greater problems stemmed from the tendency of mining

[2] Evanescent companies, which existed, at least on paper, for only a few weeks or months, filled the news columns of the major mining journals. For one example, see N. E. Guyot, "Cripple Creek: An Inside Story," *Engineering and Mining Journal-Press* 118 (December 13, 1924): 936. *EMJ* 25 (April 27, 1878): 287; Green to Samuel P. Colt, March 21, 1892, Letterbook 6, Norvin E. Green Papers, The Filson Club, Louisville, Ky.

companies to capitalize their properties at figures far above the real or sensibly estimated value of the mines. If the purpose of mining finance was to provide adequate capital for developing a mineral lode into a productive enterprise, then stock watering, as an "unhealthy, hypertrophic growth" on the industry, worked at cross purposes with that goal by frightening away the more timid investors. Cautious capitalists, fearing that a mine could never repay the principal on an investment or return a satisfactory margin of profit, shunned those companies, leaving them in the anomalous position of having a huge value on paper while they struggled for want of working capital. Usually an "ideal amount" in his mind and an unfailing faith in the project, coupled with his desire for a large promoter's fee, led the organizer of a mining company to capitalize a claim acquired for $100,000 or $200,000 in the West for a million dollars or more in New York or Philadelphia. In fact, to Frank Fossett it seemed at times that the "more worthless the lode the higher capitalization." When Boston promoters purchased the Dunkin mine at Leadville in 1879, they paid $300,000 for a property that had returned $35,000 to its owners, but it was given a nominal capitalization of $5 million in the East. Irresponsible financing of this sort strained investor confidence and plagued a multitude of companies in Colorado.[3]

The difficulties in selling inflated mining securities on a competitive stock market forced many promoters into discounting the shares, sometimes reducing their price to a mere fraction of the par value. And in view of the need of new mining companies for large doses of operating capital at the outset, this practice failed to bring in the necessary funds and caused greater consternation among stockholders, since the high paper capitalization had at least inferentially promised them enormous profits on their investment. Furthermore, these precarious financial arrangements naturally made shareholders edgy, and at the first signs of trouble in the mine—a pay streak pinching, underground flooding, and similar mechanical or natural misfortunes that temporarily halted production—they reacted by quickly trying to

[3] Dresher, *Description of Colorado*, p. 113; Frank Fossett, *Colorado, Its Gold and Silver Mines, Farms and Stock Ranges, and Health and Pleasure Resorts*, p. 575; Henry B. Clifford, *Rocks in the Road to Fortune or the Unsound Side of Mining*, p. 248; *EMJ* 22 (October 14, 1876): 246; *Mining Record* 5 (May 17, 1879): 403.

dump their stock on the market and be done with the scheme. It also became clear that overcapitalization fostered reckless handling of the shares by inviting promoters to pay themselves handsome commissions in stock and to make generous gifts to figurehead directors, who could then trade the securities or earn dividends in the same proportion as paying subscribers. The Dunderberg Mining Company at Georgetown, organized in 1879 with a nominal capitalization of $1.5 million in 150,000 shares at $10 apiece, illustrated several of these problems. Having discounted its shares by three-fifths, the company sold 63,000 of them in New York and, with the money raised, paid $242,000 for its silver claims and used the remaining $10,000 for working capital. An investigation into the company's weak financial state in 1881 revealed that 86,000 shares had been kept by the promoters and officers who received dividends along with other stockholders.[4]

"Tramp corporations" with $10 million in capital stock and only $5.75 in cash drew the ire of one western mining journal for promising more mineral production than their financial structure could deliver and for steering investor attention away from practical business matters to fanciful visions of fabulous wealth. In 1895, for instance, 126 companies incorporated in Colorado (principally to work at Cripple Creek) registered a total capital stock of $461,061,500, or approximately $50,000 more than that district produced in gold between 1890 and 1953. Fully aware of how often mining ventures utterly failed to achieve their grand designs, T. A. Rickard thought that buying shares in many of them was pure folly, and he wondered: "What share [there could be] in a mine represented by paper telling that its possessor owned so much stock in a million dollar company floated on the flimsy basis of a hungry looking pocket hole?" As knowledgeable people in the industry understood, overcapitalization was fundamentally unsound and hurt the competitive position of mining stocks on eastern money markets, especially among investors of a more conservative bent.[5]

[4] *EMJ* 25 (March 16, 1878): 195; *EMJ* 32 (December 31, 1881): 433–34; John F. Graff, *"Graybeard's" Colorado: or, Notes on the Centennial State*, pp. 72–73; Floyd Davis, *The Mine Investors' Guide*, p. 55; Clifford, *Rocks in the Road*, p. 251; John F. Hume, *The Art of Investing*, pp. 31–32.

[5] A government team reported that there were thirty Colorado mining com-

At times discussions about the ills of mining finance focused on the issue of stock assessments. Though prohibited by the corporation laws of Colorado and New York, mining companies formed in California and Nevada could assess their stockholders pro rata on the shares they owned to raise additional funds, ostensibly for development work. During the heyday of Nevada's Comstock Lode, some companies frequently treated their investors to "Irish" or "Chinese" dividends, as assessments came to be known, and, unless paid within a specified period of time, the stock could be confiscated by the company and resold. Unscrupulous operators on the inside utilized this device as a weapon in their stock manipulations, and assessments kept inefficient companies afloat, supported spendthrift company officials, and were even used sometimes to pay "dividends" to naïve stockholders. Indeed, one critic calculated that an enterprise with a paper capital of ten million dollars and a moderately poor mine could levy assessments for twenty years and probably receive one-half million dollars in small installments from credulous investors. Between 1888 and 1898, a Boston company called the Geyser, operating at Silver Cliff, Colorado, reportedly collected nearly one million dollars from its stockholders, who "courageously" endured nineteen assessments to keep their share in the mine by routinely sending good money after bad. It led a local wag to quip that if the "New England suckers" did not find silver in depth, they might at least reach the point where "blind codfish and sightless herrings are located."[6]

New York laws designed to eliminate fraud and stock tampering by prohibiting assessments, despite their good intentions, did not earn high marks from many experts on the mining industry. Paul Ward,

---

panies in the early 1880s with a total nominal capitalization of $142,100,000, but with a market value of merely $23,925,000 (Samuel F. Emmons and George F. Becker, *Statistics and Technology of the Precious Metals*, U.S. Dept. of the Interior, Census Office, p. 116). *MSP* 70 (March 14, 1895): 161; *MSP* 72 (January 11, 1896, and February 1, 1896): 23, 83; Frank Hall, *History of the State of Colorado*, 4: 106; Thomas A. Rickard, ed., *The Economics of Mining*, p. 6.

[6] In the late 1890s, Colorado made assessability optional in mining companies *MSP* 81 (July 7, 1900): 3). *Ouray Solid Muldoon* in *Financial and Mining Record* 30 (July 4, 1891): 12; Joseph G. Martin, *A Century of Finance: History of the Boston Stock and Money Markets, January, 1798, to January, 1898*, p. 230; Pope Yeatman and Edwin S. Berry, *Mining Securities*, p. 5; Frederick H. Smith, *Rocks, Minerals and Stocks*, pp. 188–89.

a respected correspondent for San Francisco's *Mining and Scientific Press*, examined New York financial practices in 1881 and concluded that nonassessability denied good mines in the West adequate capital for development and forced eastern investors into "passing around the hat" when more money was needed. He also pointed out that western merchants and mine laborers naturally hesitated to furnish goods or services to a nonassessable company, since it might not honor its obligations, thus encumbering a new enterprise with an entirely avoidable credit problem. Somewhat gratuitously he remarked that easterners no doubt saved themselves money by these laws, but had New York controlled the Comstock, it would have most likely given up on the silver lode before hitting the great bonanza. Frank Fossett, always a prudent judge of the Colorado industry, criticized the restrictions on assessments for creating "dog-in-the-manger policy" by which a few determined stockholders advanced money for exploration, development, and machinery, while the nonpayers shared equally in the rewards of production. Yet, more commonly, as he knew from first-hand experience, eastern shareholders usually fell to quarrelling among themselves over when and how much money should be sent west, while the mine sat idle and the Colorado industry suffered another setback. In one case the Alps and Granada gold mining companies of New York exhausted most of their working capital on building a fancy mill at the head of Russell Gulch in 1864 and had to suspend operations. Being unable to attract new money, the two companies, joined by an interlocking directorate, passed into the hands of a receiver, who paid off the debts between 1866 and 1877, when a New York Supreme Court order dissolved the corporations.[7]

A New York journal regarded the "nonassessable stock system to be particularly dangerous to the interests of small investors." Without assessments, companies used their limited sums of working capital to open the mines hastily and declare large dividends at once, which hurt long-term development, or found that they needed to mortgage their properties on unfavorable terms, which led to foreclosure and

[7] *Bullion* 2 (December 16, 1879): 2–3; *EMJ* 23 (May 12, 1877): 320; *EMJ* 29 (March 24, 1880): 183–84; *MSP* 43 (September 17, 1881): 186; Frank Fossett, *Colorado: A Historical, Descriptive and Statistical Work on the Rocky Mountain Gold and Silver Mining Regions*, p. 297; Fossett, *Gold and Silver Mines*, p. 289; *United States Annual Mining Review and Stock Ledger*, pp. 18–19.

loss of the mines. Of course, total collapse of an enterprise left everyone a loser, but among small investors, who had purchased the working capital after promoters and insiders had taken stock for their services, the loss was keenly felt because it often represented all or a large part of their cash savings. Demanding more than could be expected from the typical mining company, the *Engineering and Mining Journal* suggested that companies make the unassessable system work to good advantage by using their working capital conservatively and by keeping a large bank balance so that development could progress steadily, with new ore reserves being uncovered before the last were depleted. In this way an undeveloped property would receive capital in advancing stages and at no point would all of the available funds be expended in outfitting the mine until it showed pay dirt capable of replenishing company coffers. Nonassessability could then foster sound mine development as well as protect the interests of the investing public.[8]

A precious-metal mine in financial straits (particularly one that had been promoted as a sure-fire investment) was an extremely weak magnet for new money, and, with the regulations against assessments, it challenged the ingenuity of company officers to refinance the operation. In addition to seeking short-term bank loans and perhaps paying workers in promissory notes, a company might mortgage its property and ask stockholders to purchase bonds in proportion to their shares. Though the bonds were often substantially discounted, success depended on shareholders' renewing their faith in the mine's profitability, the desire to keep the company alive and solvent, and the will to shore up the value of shares already in the hands of investors. However, with uncertainty about the enterprise building, no stockholders wanted to be the first to reinvest. Discouraged in attempts to sell $6,000 in bonds to relieve his Clear Creek company "from jeopardy and put the property either in a condition for an advantageous sale or the extension of the work," Bradford Prince reported that each stockholder waited for the other, and it seemed to him a "dreadful thing that the property should be sacrificed simply because the parties most interested cannot act simultaneously." In 1880 promoters of the Big Pittsburg Consolidated Silver Mining Company at Lead-

8 *Mining Record* 6 (August 23, 1879): 141–42; *EMJ* 23 (March 17, 1877): 177–78; *EMJ* 29 (March 27, 1880): 216; *MSP* 39 (September 20, 1879): 177.

ville hoped to retire a debt by offering a large block of stock to their investors at a price much below par and even below the market value. When this backfired (only one-half of the trustees took the option), the company mortgaged its properties and issued $300,000 in bonds, valued at $500 each but sold at $50, which would have brought the company $30,000 towards paying $80,000 in debts and made it responsible for the principal at par plus 12 percent interest. Despite the urgings of promoters that "everyone should buy and thus protect his interest in the enterprise," Cyrus McCormick, a leading stockholder with a mine shaft named in his honor, saw the flaws in the scheme, refused to purchase the bonds, and instead dumped his 7,400 shares on the market. A year later the Big Pittsburg, still struggling to survive, changed its name to the New Pittsburg Mining Company and asked stockholders to trade their old shares for the new ones for an additional twenty-five cents.[9]

Such company reorganizations usually involved a slight name change, a reduction in the capital stock, and some reshuffling of the management, with the intention of consolidating or removing old debts and starting anew, ideally avoiding the mistakes of the past and exercising greater restraint in financial matters. Yet reorganization might also camouflage a raid on minority stockholders or conceal an attempt at dodging company creditors. A stockholder in the malodorous Bull-Domingo Mining Company of Silver Cliff accused the officers, among them William H. Barnum, chairman of the Democratic national executive committee, and promoter Stephen W. Dorsey, of trying to seize the company in payment for a dubious debt of $120,000. Eventually the company defaulted and its properties were purchased at a sheriff's sale by several of the trustees. In 1885–1886, when the company still could not pay thousands of dollars in back wages to its employees, a mysterious explosion rocked the mine, in-

[9] Prince tried to sell his bonds of fifty, one hundred, and five hundred dollars at 90 percent of their face value. Unsigned [Prince] letter to "Dear Sir," July 9, 1898, and J. O. Carper to Prince, November 30, 1898, LeBaron Bradford Prince Papers, New Mexico State Records and Archives Center, Santa Fe. A. L. Earle to McCormick, April 15, 1880, June 25, 1880; William M. Lent to Jesse Spalding, December 6, 1880; A. L. Earle to Jesse Spalding, December 9, 1880; A. L. Earle to McCormick, March 9, 1881; Statement of the Big Pittsburg Consolidated Mining Company, December 20, 1880; all in Series 2A, Box 34, Cyrus H. McCormick Papers, State Historical Society of Wisconsin, Madison. *EMJ* 35 (February 24, 1883): 107.

juring and killing several of the workers. Grabbing the opportunity to reorganize, the insiders turned the enterprise into the Phoenix Lead Mining Company and simply refused to recognize any claims from widows and orphans of the tragedy. All signs indicate that the mine workers and small shareholders in the Bull-Domingo paid the price in a shrewd maneuver to capture and remake the company.[10]

Since mining ventures typically evolved from a joint effort by a few friends and associates to organize and promote a company, small shareholders, who bought into the enterprise later, often found themselves at the mercy of insiders. For instance, in addition to David Moffat and his long-time partner Eben Smith, the board of directors of the Anaconda Gold Mining Company at Cripple Creek in 1899 included three men who had been employed by Moffat and Smith at various times and were used to working together. Under conditions of this kind, little stood between minor stockholders and the manipulation or shady dealings of unscrupulous directors whose management of the mine and its finances enabled them to juggle stock prices in their own interests. The normally hard-headed mining engineer Eben Olcott, a small stockholder in the Silver Cliff Mining Company, frankly admitted that insiders would make "lots of money" by manipulating the stock of his company, but he confessed he knew "nothing of such operations & you can't tell which way they will move unless you are in the ring." Those on the outside and, in the case of eastern investors, those far distant from the property as well, could easily be misled and gulled by the controlling cliques, whose techniques included paying dividends from loans or the sale of securities, freezing out the small owners, or indeed doing most anything the imagination conceived, as long as it remained hidden. Mining deceits, conceded one expert, showed a "brilliancy of execution and a clearness of cut in the pattern, marking them among frauds in general with a degree of distinction which attracts the attention at once."[11]

[10] Fossett, *Gold and Silver Mines*, pp. 478–79; *EMJ* 30 (July 31, 1880): 13; *EMJ* 33 (March 11, 1882): 135; *EMJ* 35 (January 20, 1883): 33; *EMJ* 41 (February 27, 1886): 156; *EMJ* 43 (March 5, 1887): 173.

[11] *EMJ* 23 (January 20, 1877): 38–39; *EMJ* 67 (February 11, 1899): 183; Eben Olcott to "Euric," July 17, 1881, Ebenezer E. Olcott Papers, New York Historical Society, New York City; Clifford, *Rocks in the Road*, p. 155; William R. Balch, comp., *The Mines, Miners and Mining Interests of the United States in 1882*, p. 859.

Free of legal compulsion, mining companies generally did not issue regular itemized statements of profit and loss that might have better informed and guided stockholders in their decision making, especially where the efficiency of management or the future of the enterprise was concerned. In part, this negligence can be explained by the scarcity of qualified bookkeepers, the ignorance of mine clerks in the methods of accounting, and, perhaps more so, by the casual attitude toward financial matters that characterized so many companies, notably the smaller ones, operating in a boom and bust industry. But nineteenth-century businessmen also stoutly defended the view that companies were private affairs closed to the eyes of the curious and duty-bound to keep the details of input-output, production, bank balances, earnings, and the like from competitors, who might gain an advantage from something contained in the company report. Ironically, while western mining companies refused to disclose their financial conditions, some promoters of the industry falsely claimed that gold and silver mines, in sharp contrast with other businesses, did not compete, and that precious metal, no matter how much was produced, would always find a ready market at good prices. When asked about his reasons for not making public information about his operation, the president of one large mining company replied: "It is our private business, and why should we give it to the world to satisfy the curious, or help educate young mining engineers who have not had the practical experience along those lines, or to furnish ammunition for the stockholders to make erroneous comparisons between two mines of perhaps very different type?" British mining engineer Philip Argall added that some mining men in the Rockies feared that itemized reports would be used by railroads and smelters to raise the rates on transportation and ore reduction.[12]

The major mining journals encouraged companies to release some financial information at regular intervals in the best interests of the whole industry. In 1880 the *Mining and Scientific Press* saw "no good reason" why mine owners should withhold the facts and figures on cost and production, at least in general terms, since they would instill a sense of competition between companies and lead to some

[12] Company president quoted by Argall in Rickard, ed., *Economics of Mining*, p. 300; Balch, comp., *The Mines . . . in 1882*, p. 863; *EMJ* 29 (January 24, 1880): 61; R. H. Stretch, *Prospecting, Locating and Valuing Mines*, pp. 2–3.

good-natured rivalry for the sport in it. Agreeing that only "obtrusive intermeddlers" would want to poke their noses into company ledger books, it thought that broad statements indicating "success" would foster further exploration; attract more workingmen and merchants into the West, thereby reducing the cost of wages and supplies; increase the industry's knowledge of itself; and satisfy the public's right and desire to know how the mines were progressing. The *Engineering and Mining Journal,* which had echoed these views on several occasions, rejected the following "excuses" in 1887 for not presenting full and frank statements on financial matters: the general public has no right to know and shareholders can apply to the company privately; mine managers are honest men who should enjoy the complete confidence of investors; company directors own large blocks of stock and so will run the firms wisely; and the cost of production depends on trade secrets. Angry at the large number of enterprises that never furnished a report from year to year, the same paper eventually advocated monthly financial statements as the fairest treatment for stockholders and also as the surest way of proving to the money centers of the East the honor and legitimacy of mining.[13]

When reports did issue from the companies—sometimes at the behest of stock exchanges—they were usually vague and evasive or crafty pastiches of bluff, blandishment, and laundered figures aimed more at boosting stock than giving a straightforward account of financial conditions. Classified by some as "literature of fiction," company reports hoped to keep stockholders satisfied with "sanguine and unreliable predictions of profit" in the near future. One critic especially deplored the habit of mining companies claiming profits when short-term earnings surpassed expenses, which could easily mislead investors since few mines maintained a sinking fund to amortize the large initial expenditures on the purchase of property, reduction works, and heavy equipment. Stockholders lulled into thinking that dividends equaled profits might awake one day to find that those dividends never repaid the principal on their investment. Depending solely on

---

[13] When *Bullion* proposed that federal government examiners visit the mines and issue regular reports, Rossiter Raymond denounced the idea as "unconstitutional, impertinent and tyrannical" (1 [May 1, 1879]: 1–3). *MSP* 40 (February 7, 1880): 88; *EMJ* 43 (April 9, 1887): 254; *EMJ* 52 (November 14, 1891): 561; *Mining Record* 7 (January 24, 1880): 75.

mine officials for accurate information on the wisdom and value of company operations invited fraud and led many investors to ruin.[14]

However, if minority stockholders demanded the right to examine company ledgers or even tried to visit the mine, they often ran into stubborn resistance from the officers, who, besides fearing the spread of stories harmful to the sale of stock, objected to greenhorns inspecting what they could not understand, getting in the way of working miners, and generally making a nuisance of themselves by asking foolish questions and stumbling around the open mine shafts. Engineer Theodore B. Comstock found that mine managers in the San Juan region despised small investors as "confirmed grumblers," who caused the most trouble over a $100 investment. Yet under New York State law the owner of 5 percent of the capital stock in a company not exceeding $100,000 or 8 percent in one over $100,000 could request a written report from the company treasurer, who was then bound to give the true facts under oath, and failure to comply within twenty days constituted an offense punishable by a $50 fine and $10 for every additional day. These reports written under pressure proving unsatisfactory and evasive, some of the bolder shareholders decided to see for themselves and headed for the mine, only to find that management prohibited visitors and had sworn all employees to absolute secrecy. This was the case at the Mollie Gibson mine at Aspen in 1892 and 1893, when Judge Moses Hallett of the United States District Court in Denver ruled that a stockholder had the right to inspect the workings, even though the mine had paid $1.8 million in dividends over the previous twelve months. Furthermore, the introduction in the Colorado legislature of bills that would grant an unlimited right of examination to investors or set aside a prescribed period of time each month for visits by those holding 1 percent or more of the stock aroused vigorous opposition from the mine owners. In fact, the son of James J. Hagerman, president of the Mollie Gibson Consolidated Mining and Milling Company, called visits by minority stockholders "very vexatious," and reported that his father spent $500 of company money to "kill" one bill before the state legislature in 1895. By shrouding their operations in secrecy,

[14] *EMJ* 23 (February 3, 1877): 69–70; *EMJ* 29 (March 27, 1880): 215; Floyd Davis, *Mine Investors' Guide*, p. 59; James R. Finlay, *The Cost of Mining*, p. 15; Seth G. Pope to Prince, July 13, 1895, Prince Papers.

company officers added another seemingly unnecessary obstacle to investor confidence in the western mining industry.[15]

Spared close scrutiny by absentee capitalists and infected by the boundless optimism that spread during the boom days, mine management usually became careless and extravagant. As Raymond once observed, "the mistakes of mining are always to a greater or less extent irretrievable," and, unless management practiced strict economy and planned development soundly, the finite body of ore within a mine would be depleted before shareholders had earned a fair return on their investment. That fragile balance between the amount of money expended on the operation and the mine's rate of decline from a rich beginning to economic exhaustion had to be maintained and, should the balance be lost, the whole enterprise teetered on the brink of financial disaster. Few experts disagreed with the outline for success in deep-level mining presented by one journal in the 1860s:

We cannot see how any great loss can be made where a good substantial mine is taken as the basis; where sufficient amount of capital is embarked in the enterprise; and where that capital is judiciously expended with prudent care and wise forethought upon the legitimate development of the property.

Yet poor management of the mines due to ignorance and inexperience hobbled the industry in Colorado throughout its first decades.[16]

Some critics roundly condemned the loose-spending company officers in the East who, in their nearly always premature exultation that the mines had made them millionaires, squandered the unaudited funds of shareholders on high salaries, plush offices, and promoters' fees, and surrounded themselves with a retinue of ne'er-do-well

[15] In 1887, the director of the Mint reported that he never solicited cost sheets from mining companies, "nor is it likely that many would be forthcoming however desirable or whatever manner sought. . . . Public inquiries into cost of production of almost any commodity are repelled by producers and manufacturers" (U.S. Bureau of the Mint, *Report of the Director of the Mint upon Production of the Precious Metals in the United States during the Calendar Year 1887*, pp. 89–90). *EMJ* 32 (October 29, 1881): 284; *EMJ* 55 (January 14, 1893): 37; *MSP* 43 (September 10, 1881): 175; *Rocky Mountain News*, February 3, 1885; Percy Hagerman to David H. Moffat, April 3, 1895, Box 1, Eben Smith Papers, Denver Public Library.

[16] Rossiter W. Raymond, *Mineral Resources of the States and Territories West of the Rocky Mountains*, p. 176; *AJM* 1 (April 7, 1866): 24.

friends and relatives generously supported from the company treasury. These "kidglove humbugs," charged the *American Journal of Mining* in 1866, sat in their "elegantly furnished offices, rolling back in easy chairs, with their heels cocked up on the window sill or table, dressed in the latest cut, lazily puffing fragrant Havanas, insolent to all callers save officers and heavy stockholders, and evidently firm in the belief that their sprouting little acorns are already enormous oaks." At the same time that top-heavy management in New York or Boston spent money recklessly on nonessentials, some companies tried to hold down costs by cutting wages at the mine, hiring inexperienced or incompetent superintendents, or skimping on tools and equipment needed for good development work. Economics "practiced in the small things and neglected in the greater" became ludicrous when such a company reduced miners' wages by fifty cents a day by order of a $5,000 president in a $2,000 eastern office through a $2,500 secretary, even though stockholders might be "charmed" by these efforts to reduce the expense of production.[17]

Hurrying to begin work on a newly discovered vein of ore, and ignorant of the real difficulties associated with hardrock mining on the frontier, absentee companies often selected totally untrained and unfit men as superintendents and entrusted them with the day-to-day management of the property, the coordination of men and machines, and the purchase of supplies. By choosing a novice to bear these heavy responsibilities—sometimes on the thin basis that he was available for the job, had a nodding acquaintance with machinery, or enjoyed the recommendation of friends—mine owners no doubt took refuge behind the old saw that not even poor management could ruin a good mine. Yet that proved to be misplaced confidence as many companies sent out inept managers from the East who damaged their mines through inexperience, incompetence, and negligence, and earned epithets from the western press. An Idaho Springs newspaper in 1887 created a typical "dude" superintendent, named him "H. Algernon Sidney," and explained that he had received his $5,000-a-year job because of his relation to a "big director," even though he had never seen a mine before or worked a day in his life. Naturally he made a trifling contribution to the business of mining in Colorado; besides

17 *AJM* 1 (September 1, 1866): 360; *MSP* 41 (October 9, 1880): 232.

an extensive wardrobe of five dress shirts and collars and cuffs, he brought along articles for his toilet, a fifty-dollar fishing rod, a handsome rifle, his favorite billiard cue, a bull dog, and a trunk full of novels to " 'kill time in the beastly place.' " Since mines could be found more easily than qualified superintendents, the Rocky Mountain industry employed a wide array of untrained managers, from shopkeepers and tailors to schoolteachers gone West to make a fortune. The Smith and Parmelee Company boasted in 1869 that it had retained a "practical businessman of long experience"—the former chief mechanic of the Milwaukee and St. Paul Railroad—to run its mine, and in the 1870s the manager of the Russia mine at Alma had previously been a physician in Indiana. Placed in the position of operating a mine, which required more skill and experience than running a grocery or a small factory, the "ordinary salaried mining superintendent," said an investment counselor, was good only to "draw checks on his employers, and bury their money beyond the reach of resurrection."[18]

Acknowledging that each mine had its own distinctive geology and that each company wrestled with individual problems due to locale, labor supply, transportation, and ore reduction, experts cited a few generally applicable rules of conduct aimed at improving mine management: sound and careful engineering; using men and machinery efficiently; practicing economy in the purchase and consumption of supplies; and holding in reserve a portion of the high-grade ore to meet unexpected and extraordinary expenses. These rules were grouped together as a basis for the "businesslike" approach to mining for precious metals. Because gold and silver mines usually contained their treasure in short and isolated shoots of ore, which could pinch out suddenly and necessitate weeks or even months of "dead work" in the worthless surrounding rock before relocating the vein, it was imperative that companies work conservatively and maintain a cash reserve or a block of pay ore in the mine in the event of an emer-

---

[18] One mine owner stated flatly: "It is easier to get a mine than a man [manager]. Give me a man and I will get the mine" (Francis C. Nicholas, *Mining Investments and How to Judge Them*, p. 189). *Idaho Springs* (Colo.) *News* quoted in *MSP* 55 (November 5, 1887): 291; Harry J. Newton, *Pitfalls of Mining Finance*, p. 63; Hall, *History*, 3: 441–42; *AJM* 7 (February 13, 1869); 99; Hume, *Art of Investing*, p. 112.

gency. With no reserve capital, some companies faced premature collapse (more so if unable to make assessments), while querulous shareholders debated the wisdom of investing more money at this gloomiest stage in a mine's life.[19]

The irregularities that occurred in most ore formations gave conservative companies greater reason to leave some ore in place as a reserve, which would not be removed until new pockets of ore had been discovered and measured, and made steady and orderly development work appear as the best means of periodically renewing the life of a mine until the deposit was thoroughly exhausted. Sound managerial practice also dictated that dividends to shareholders be withheld until the company treasury contained ample sums of money to furnish ready cash in the event of a labor dispute, fire, or flooding, and that a sinking fund be established to repay the full amount of invested capital. Mining engineer Herbert Hoover, a brilliant practitioner of his craft and a perceptive critic of mine management, stressed the point in 1909 that a mine could not claim success merely by declaring higher returns than expenses, but needed to set aside a portion of its earnings each year to redeem the original investment made by financiers. Unless an investor understood the "ephemeral" nature of a mine, he stood to lose the principal on his investment, although he received dividends temporarily. The *Mining and Scientific Press* in 1898 recommended an ore reserve in the ground or money in the bank as the surest protection against the risks and short life of a mine and warned against management declaring premature dividends, robbing a mine for current expenses, or shipping ore too quickly to impress the public. Whether from "crime or folly," reported the journal, "nearly" every prominent mining camp in the Rockies had repeated examples of companies depleting usable ore reserves and abandoning mines with pay ore still locked in their depths. Consequently, "the argument will always be, if a "property is in pay, it can maintain a reserve fund, and if it cannot maintain a reserve fund, it cannot pay."[20]

19 *EMJ* 8 (August 10, 1869): 83; Herbert C. Hoover, *Principles of Mining*, p. 161; J. C. Pickering, *Engineering Analysis of a Mining Share*, p. 31.

20 Actually Hoover thought mining companies would seldom establish a financial system to amortize invested capital, but he offered this idea mostly for the benefit of investors who could then judge the risks for themselves (*Principles of*

Though extraordinarily wasteful and shortsighted, most Colorado mines in the boom years prior to 1900 led a "short life, but a merry one" in the hands of reckless and inept managers. Considering the scores of amateurs directing mining operations in the Rockies, it is little wonder that slower, sounder techniques based on strict economy and sensible development of the ore deposits foundered and that the urge to strip a mine of its richest mineral prevailed. The mining engineers who advocated cost consciousness and conservative business methods butted heads with managers who felt that it was "poor business to work for posterity," when the great western frontier seemed to offer unlimited opportunities to exploit its natural wealth. Whatever the cause, many managers ruthlessly gouged and robbed the mines and in the process badly damaged or permanently destroyed good properties, making valuable ores inaccessible and stranding the owners on a sea of worthless stock in the enterprise. Soon after the Empress Mining Company took over a partially developed mine on Mammoth Hill near Central City in 1879, the new superintendent timidly reported to eastern owners, among them the mayor of Brooklyn and the police commissioner of New York City, that "promiscuous gophers" in the previous management had removed hundreds of tons of the highest-grade ore.[21]

Wonderful tales about the richness of the western mines and the glib promises of instant wealth peddled by aggressive promoters aroused the cupidity of eastern investors and produced strong pressure on new and weak companies, uncertain about the proper course for development, to remove the richest ore bodies at once. If the management failed to return large and speedy dividends, shareholders became anxious and began prodding the company, sometimes threatening law suits or public disclosures of wrongdoing by the officers. On at least one occasion, shareholders resorted to violence: the Cleveland police were summoned in 1892 to halt a brawl between the directors and stockholders of the Magna Charta Silver Mining Company over

---

Mining, pp. 22, 42–44, 189). Clifford, *Rocks in the Road*, p. 140; Newton, *Pitfalls of Mining Finance*, pp. 94–95; *MSP* 76 (June 25, 1898): 663.

[21] Rickard, ed., *Economics of Mining*, pp. 158–59; *EMJ* 28 (August 9, 1879): 100; Fossett, *Gold and Silver Mines*, pp. 318–19; Joseph W. Holman to "Gents," March 22, 1879, Empress Mining Company Papers, Western Historical Collections, University of Colorado, Boulder.

low returns and high expenses. A better illustration of stockholder discontent appeared in the correspondence between Devillo R. Holt, a prosperous Chicago lumber and real-estate operator, and his son George, who managed the Little Chief Mine in Leadville, owned principally by Chicago merchant John V. Farwell. As well as having loaned George the money to buy Little Chief stock, Devillo was friendly with several of the largest shareholders. He kept a close watch on the development of the company and secretly reported to his son the changing plans and moods of the owners. In March, 1879, for instance, he wrote that all of the Chicago men were expecting to be millionaires "from the mine in a short time," and the following month he added that they were "all crazy about the mine and expect a constant stream of silver running into their pockets, and if they do not get it, will lay the blame upon the management." Indeed, when the flow of ore from the mine did not meet expectations, stockholders began to scold George and insist upon dividends at once. After one agitated meeting of the Chicago owners, Holt counseled his son: "I do not see any way of remedying them but one, and that is by sending along big dividends, nothing else will have any weight, making explanations and bringing charges against others no matter how just will never give the money and money is what they want." Eventually, under great pressure, George resigned and, in defending his management, declared that he had faithfully followed the instructions of company officers to " '*take out big money even at a loss*' " in order to satisfy impatient stockholders. In that, he felt "we have done as well as we could" within the limits of his own conviction that the mine should not be sacrificed to haste and greed. Still, investors continued to conspire against their own best interests by demanding quick dividends, which usually ended in ruin for the enterprise and a sharply taught lesson in the waste of exploitative operations.[22]

As George Holt and many other agents of eastern companies discovered to their chagrin, the hundreds of miles separating them from the home office made communications very difficult and aggravated

[22] *EMJ* 53 (February 6, 1892): 187. D. R. Holt to George Holt, March 13, 1879, April 25, 1879, June 28, 1879, Folder 39; George Holt to Executive Committee of the Little Chief, undated, Folder 40; all in George Holt Papers, Western Historical Collections, University of Colorado, Boulder. On Devillo Holt, see George W. Hotchkiss, *Industrial Chicago*, vol. 5, *The Lumber Interests*, pp. 315–19.

nearly every disagreement over plans or purposes of the management. Although railroad and telegraph services steadily improved in the West during the last third of the nineteenth century, major problems persisted. Mail delivery was slow and unreliable, particularly into the isolated mining camps, and weekly telegrams meant to give details on development work or explanations of the finer points in mining geology were too expensive and failed to satisfy information-starved shareholders. Even when the physical problems of communication could be surmounted, there remained a more basic gap between East and West: the chasm between the high expectation of absentee capitalists and the unpleasant realities of mining as reported by the superintendent. If the news from Colorado failed to improve over a period of time, easterners often suspected the worst of their managers, supposing everything from intrigue to incompetence, with the result that further, serious communication of the facts became well-nigh impossible. At such times, with his good will and common sense being questioned, the mine manager needed to be a master diplomat to present his side of the issue without rancor, ask tactfully for more time or money, and smooth down the ruffled feathers of stockholders, while rebuilding trust in his management. On this score Holt and many other superintendents failed. Holt bristled at charges of poor management and became angry and argumentative when several Chicago investors accused him of spending their money too freely, working too little, keeping more men above ground than in the mine, and hiring an assistant who did little more than " 'fumb' around & fume & curse." Asserting that he personally supervised operations at the mine for sixteen to twenty hours each day, Holt rejected his father's advice to "meet the trouble with a firm purpose & leave all results with God," tendered his resignation, and headed home.[23]

For several months in 1879 the superintendent of the Empress mine at Leadville, Joseph Holman, filed effusive reports with his New York employers expressing complete confidence that the "mine will prove highly satisfactory to all of you and prove a good paying mine at no very distant day." Despite still more earnest messages about the "cistimatic order" he had instituted at the mine and the progress made

---

[23] D. R. Holt to George Holt, May 5, 1879, May 6, 1879, Folder 39; George Holt to Executive Committee of the Little Chief, undated, Folder 40; all in Holt Papers.

under his "good stewardship," the owners wondered why nearly $20,000 had been spent and no profits had been sent their way. One inquisitive shareholder visited Leadville and consequently charged Holman with sending deceptive reports, doing only half the work he claimed, and, in short, having *"Beat us Bad"* by pocketing over $5,000 in company funds. Whether Holman had indeed swindled the enterprise is hard to say, but it is clear that distance vexed every effort to coordinate management of the western mines and at the same time gave greater license to skulduggery at the expense of absentee investors. Besides complicating the duties of management, it also bred crises in confidence, involving real and suspected misdeeds, within an industry that already contained more combustible elements than most and that needed the fullest measure of integrity and square dealing to shed the image of pure speculation.[24]

Larger capitalists who could not make the long and arduous trip west to examine the mines personally, but who also wanted regular and reliable reports from the field, frequently employed special agents in Colorado to notify them of newly discovered lodes or attractive mining proposals as well as to watch over their previous acquisitions. If the relationship functioned well and a solid sense of trust and understanding developed, the absentee investor benefited from having his own eyes and ears in the mining regions, and he usually empowered his agent to act in his behalf under specified conditions. With harmony and trust essential to this arrangement, investors also sought men who possessed good common sense, could read and write, and desired to learn mining or had a working knowledge of its fundamentals. As generally the best-educated group of people on the frontier, local lawyers satisfied many of these qualifications, and, in addition, their knowledge of the law was valuable in handling a variety of legal questions, from preparing company documents to directing litigation. New York merchant Jerome B. Wheeler, a major investor at Aspen, transacted much of his Colorado business through three agents, each from a different background: George J. Boal, once a lecturer in medical jurisprudence at an Iowa school, a vice-president of the American Bar Association, and a Democratic candidate for

---

[24] Joseph Holman to James Howell, April 15, 1879; Holman to "Gents," March 29, 1879, August 17, 1879; Junius Schenck to Peter French, August 1, 1879, August 2, 1879, August 8, 1882; all in Empress Mining Company Papers.

governor of Iowa; W. W. Cooley, a shrewd country lawyer from Kansas; and Henry R. Woodward, a Harvard-educated salesman whom Wheeler met casually while vacationing in Virginia. In the late 1880s and 1890s, Captain Lafayette E. Campbell, who had previously supervised construction of Fort Logan in Denver, watched over the interests of millionaire David Moffat in the mines around Creede. Finally, the investments of Abel D. Breed of Cincinnati bore the stamp of his chief agent Henry B. Gillespie, an honor graduate from college, who had been a passenger agent for the Santa Fe Railroad prior to his employment by Breed at the Caribou mine in the 1870s. For obvious reasons, the representative of a wealthy businessman operated most effectively when he could keep the name of his employer secret, and many agents did this job so well that they remained obscure figures themselves, always present in the Colorado mining camps but rarely emerging from the shadows of hearsay. On the whole, however, having confidential advisers in the West helped chart the course for absentee capitalists in their decision making and provided them with rapid and personal appraisals of the fickle business of mining.[25]

A growing number of investors after the Civil War also realized that the hazards and complexity of deep-level mining, along with the vast sums of capital involved, required expert advice and technical know-how, and for them the trained mining engineer became an indispensable ally, though finding a genuine and capable one often proved difficult and carried its own risks. Throughout the era, quacks and charlatans—most of whom literally could not tell the difference between ore and porphyry—proclaimed themselves "mining experts," tagged M.E. onto their names, and paraded before the gullible public as bona fide engineers. Their strongest point, according to one journal, was the capacity to judge "ore in sight," by which they "loosened the purse strings of many a doubting Thomas among eastern capital-

[25] D. Bauman, *King Carbonate, Leadville, Colorado*, p. 54; Hall, *History*, 4: 609; John G. Canfield, *Mines and Mining Men of Colorado*, p. 40; *Portrait and Biographical Record of Denver and Vicinity*, pp. 655–56; John J. Lipsey, *The Lives of James John Hagerman*, pp. 125–26; Robert Pfanner, "Highlights in the History of Fort Logan," *Colorado Magazine* 19 (May, 1942): 83–84; Lafayette Campbell to Eben Smith, October 7, 1889, March 28, 1891, Box 1, Smith Papers.

ists." Especially during a mining craze, the self-styled mining expert found a ready market for his artifices in the eastern cities.

He has himself interviewed by the mining papers and talks knowingly of shafts and winzes and stopes and hoisting works and stamps and amalgamating pans and formations and gangues and lodes and veins and seams and faults, and manages to make the interviewer think him the man who built the Comstock lode. Some of these experts talk as if they could put ore in a barren mine, or make a granite boulder assay ten thousand dollars a ton.

Occasionally the pretense was punctured, as happened to one "professor" who discovered gold and silver in a crushed jug handle, inspiring pranksters to ask him whether a dried herring came from a "true fish-er vein" or a pile of manure indicated "horse carbonates in place."[26]

Imposters aside, trained mining engineers had to put to rest some doubts about their own worth and integrity before they could earn the confidence of financiers and make their considerable talents felt in the industry. To start with, ordinary mining men in the West scorned the fancy, educated fellows from eastern and foreign schools who knew a great deal about the laws of science, but who supposedly did not possess the practical knowledge to judge a mine's value, develop it, or manage the men and money that made a gold or silver streak pay. In 1883 a Clear Creek miner charged that the young graduate engineers fresh out of school and weighed down with honors from Freiberg or the Columbia School of Mines were "all theory—no practice—no business education." Nor could they find ore with their learned theories, according to the joke circulating in Colorado in the 1890's: "How are things in Cripple Creek?" "Oh, they are about the same as usual. The tenderfeet are taking the ore out where they find it, and the mining engineers are hunting for it where it ought to be." However, serious commentators realized that the mining schools

[26] "Self-made men are all very well, but not self-made mining engineers," warned the *MSP* (45 [August 5, 1882]: 88). *MSP* 40 (May 15, 1880): 312; *MSP* 41 (August 21, 1880): 120; Clifford, *Rocks in the Road*, pp. 214–16; Miguel A. Otero, *My Life on the Frontier, 1864–1882*, pp. 228–29. For a thorough examination of the professional engineer in the West, see Clark C. Spence, *Mining Engineers & The American West: The Lace-Boot Brigade, 1849–1933.*

did not produce finished engineers, that field work added the necessary seasoning, and that the real expert was the one who learned "his
lesson by studying the accumulated knowledge of all who have gone
before him, and handling the pick, gad, hammer and drill in shaft
and drift and stope, and who combines with such knowledge that
rarest of all gifts—common sense."[27]

Called upon to examine a mining venture for distant capitalists,
it was charged, the mining engineer could not resist taking a fee from
both the promoter and his employers and issuing a favorable report
to the higher bidder, his leathery conscience readily yielding to his
sensitive pocket nerve. Voicing this suspicion of double dealing by
trained experts, the *Silver Plume Silver Standard* denounced the engineer who would examine a mine for an easterner at the same time he
agreed to " 'stand in' for a percent of the proceeds should the sale go
through," realizing, of course, that a recommendation was "equiva-
lent" to a sale. A local bard in the same newspaper added his criticism
of the engineer:

> He has few scruples but takes many drams,
>   and he'll take a double fee
> From the man who buys and the man who sells;
>   and serves them both loyally.

T. A. Rickard recalled that an officer of the Enterprise mine at Rico
once offered him $1,000 to report a secondary enrichment in the
property as part of a stock-market manipulation and, though refusing
to cooperate, he thought the company could get such a dishonest statement from a struggling engineer in Denver and for a smaller price.
Another engineer felt that the profession attracted a "very virtuous
lot of men, yet on the average . . . we are no better, neither are we
worse, than the ordinary run of humanity." Since "low prospective
costs, and exaggerated expectations, have started many a mining enterprise on a series of adversities," these were the greatest evils to him,
and he believed they occurred most often when an expert desperately

[27] Cripple Creek story related by Spence, *Mining Engineers*, p. 71. *MSP* 20
(May 28, 1870): 360; *MSP* 53 (July 17, 1886): 34; *EMJ* 36 (September 15,
1883): 169.

wanted work and, "under the stress of such circumstances," hid behind the uncertainties of mining and underestimated expenses. On the other hand, some people worried about the engineer who, awed by his responsibilities to the investor and too conscious of the large amounts of money involved, tended to underestimate the value of all mining properties; these "safe men" and "mine cowards," complained Henry Clifford, "rejected everything to preserve their reputation," even if they harmed the industry in the process. Financier Norvin Green called James D. Hague, one of America's foremost engineers, a "bear on all mines," and asked that he be "dispatched quietly" to evaluate the Pelican-Dives mine and "his report made to us, so that we may see it before deciding on its publication."[28]

Despite the honest errors of many and the disreputable acts of a few, the professional engineer brought much needed scientific knowledge to the industry and, in the opinion of one journal, "killed more swindles and caused the opening up of more good mines than any other class of miners." Many of the best students from the finest mining schools in the United States and Europe spent some time in Colorado, attracted by the challenge of the hardrock mines, with their vast wealth, and the high fees paid by capitalists who found their advice unrivaled for completeness and accuracy. In the rough and tumble business of mining, the entrepreneur had no sturdier and more reliable aide than the trained engineer, and, on the whole, these professionals in lace-boots performed well. They simplified the complexities of mining, from geology to square-set timbering, translated ore shoots and gangue into terms of profit and loss for investors, and slogged through the sump pits of a Rocky Mountain mine one day so that next week they could sit down in an elegant New York or Boston restaurant and discuss the wisdom of an investment with a few swivel-chair miners. One engineer jokingly compared the purchase of

28 "Observer," "The Mining Expert," *Silver Plume* (Colo.) *Silver Standard*, September 10, 1887; Clifford, *Rocks in the Road*, pp. 214–16; Francis C. Nicholas, "The Wrongs and Opportunities in Mining Investments," *Annals of the American Academy of Political and Social Science* 35 (1910): 208; Thomas A. Rickard, *Retrospect: An Autobiography*, p. 65; Peter H. Van Diest, "On the Estimation of the Capital Requisite for Investment in Mining Properties," *Proceedings of the Colorado Scientific Society* 1 (1883–1884): 61–62; Green to Theodore N. Vail, April 21, 1880, Letterbook 1, Green Papers.

a mine with buying a pig in the poke, and his job was to provide as much information "as possible about the pig and reduce the mystery of the poke to its lowest terms."[29]

Talented experts in the "secret art of mining" were expensive, and to a degree this limited their services to wealthy individual capitalists and syndicates that could draw upon plentiful treasuries. Though respected engineers sometimes demanded several thousand dollars for one report, fees never reached the heights suggested by one mine owner who explained why there were some mines with roads cut deeper than their shafts: " 'You know here in Colorado we can't get an engineer to visit our mines unless we haul him up in a buggy.' " Eben Olcott told a Brooklyn businessman in 1881 that he would perform a "thorough" examination of a claim at Animas Forks for five hundred dollars plus expenses, his report to include the following: an estimate of ore in sight; a sampling of the ore content; analyses of the cost of mining, freighting, and smelting; and a forecast of the future value and size of the ore body. Diplomatically—and free of charge—Olcott told him to forget about the claim: "There are generally one or two good properties in each permanent camp & a whole army of worthless ones that are floated on the reputation of the single ones." In choosing an engineer to examine his Pelican-Dives, Norvin Green expected James D. Hague to ask $5,000 for his fee, but, by playing coy and showing "no anxiety" for Hague in particular, Green hoped to employ him for $2,500 or $3,000. During the 1860s, Cyrus McCormick received much of his advice on mining investments from Henry A. Ward. Educated at Harvard and the École des Mines in Paris, Ward, although more devoted to paleontology and building fossil museums, worked as a mining engineer in Montana and South Carolina where McCormick invested. When McCormick politely refused to finance his museum projects, Ward offered his services to the Chicago capitalist as an engineer, and an expensive one at that. In 1868, for instance, Ward, figuring that he "must not underbid" himself, proposed to examine some Colorado mines for McCormick at a fee of $1,100—$500 for expenses and $600 for his salary—and "nothing less." His time would start as soon as he left Rochester, New York, and terminate

---

[29] *MSP* 73 (November 28, 1896): 437; Charles M. Dobson, "Mine Salting," *Cosmopolitan* 24 (April, 1898): 576; Rickard, *Retrospect*, p. 81.

when he returned to the city. Apparently girding himself for McCormick's anger at the large fee, the professor at the University of Rochester reasoned that "a lawyer who serves his client by helping him to keep out of the law, or a physician who tries to prevent sickness, or a mining engineer who seeks to stand between his employer and all delusive schemes—even though at the expense of business for himself—should not be dealt with too closely as to their terms. . . . " In spite of Ward's plan to spend a few months in Colorado, where he intended to purchase mines for 25 to 50 percent less than asked by promoters in the East, McCormick refused to spend the $3,000 the prospecting trip would most likely cost.[30]

Expensive professional advice did not end with the technical assistance supplied by mining engineers on the purchase and development of mineral deposits. For a multitude of investors, the success of their venture in Colorado depended upon the talents of men trained in the law, who could interpret state and federal mining statutes and guide the enterprise through all manner of legal snares. Lawyers helped in perfecting the title to claims, ironing out disputes over boundary lines, and fighting routine legal battles in the courts, but they had a focal role in the legal mischief caused by the federal mining law of 1872. Evolving out of the deficient federal law of 1866, and recognizing local mining customs and rules, this act, which still largely controls access to government mineral lands, attempted to "promote the development of the mining resources of the United States" and protect legitimate mine locations by granting to the possessor of the apex, or the point on a lode nearest the surface, new and extraordinary privileges to follow the vein, with "its dips, angles, and variations," outside his own sidelines and into neighboring properties. Engineer James D. Hague saw the "law of the apex" as no less

---

[30] Few investors were as fortunate as New York banker Henry Amy. As president of the San Juan Smelting and Mining Company, he appointed his son, Ernest J. H. Amy, a graduate of the Columbia School of Mines, manager of his operations (Canfield, *Men of Colorado*, p. 75). Thomas A. Rickard, ed., *Interviews with Mining Engineers*, p. 113; Olcott to M. H. Hagerty, April 20, 1881, Olcott Papers; Green to Theodore N. Vail, April 21, 1880, Letterbook 1, Green Papers. Ward to J. N. Hayes, September 12, 1868; Ward to McCormick, September 16, 1868, September 22, 1868; all in Series 2A, Box 34, McCormick Papers. William T. Hutchinson, *Cyrus Hall McCormick*, 2: 168; Roswell Ward, *Henry A. Ward: Museum Builder to America*, pp. 134, 141–54.

than an application of the Scriptural lesson: " 'To him that hath (the apex) shall more be given; but from him that hath not (the apex), shall be taken away even that which he hath.' " Furthermore, when Congress did not or could not precisely define what it meant by the apex—a term unfamiliar to miners—and left it to the judgment of the courts, the act invited huge and complex law suits to determine the apex and the ownership of a vein, while at the same time the vagueness of the statute and the endless variety of geological formations ruled out speedy and satisfactory decisions.[31]

The richness of the deep-level mines in Colorado, coupled with the zest Americans often showed for litigation, turned the state into a battleground for testing the apex law, and few good properties were able to escape at least one challenge from an adverse claimant after 1872. Rocky Mountain miners quipped that the surest way of discovering a bonanza was to check the court records, and other local wits insisted that if a mine was not in pay dirt and was not in barren ground then it must certainly be in litigation. Actually these witticisms reflected problems that had profound repercussions on the industry, for long court cases drained many companies of the time, money, and energy that might have been better spent on systematic and sound development of the mines. Because it contained high-grade ore and because the courts could not adequately define the apex, the Colorado Central mine near Georgetown faced one legal contest after another in territorial, state, and federal courts—and finally in the U.S. Supreme Court—from the early 1870s into the twentieth century. The throng of lawyers retained for these cases, as well as for those of shorter duration, convinced many people that the law enabled the miners to work the mines less well than it encouraged the lawyers to work the miners, and with some justification a valuable mine was frequently described as a "lawyer's pit."[32]

[31] Rodman W. Paul, *Mining Frontiers of the Far West, 1848–1880*, pp. 171–74; U.S., *Statutes at Large* 14 (July 26, 1866): 251–53, and 17 (May 10, 1872): 91–96; Rossiter W. Raymond, "The Law of the Apex," *Transactions of the American Institute of Mining Engineers* 12 (1883–1884): 387–93; James D. Hague, "Mining Engineering and Mining Law," *EMJ* 78 (October 20, 1904): 629.

[32] *MSP* 13 (December 15, 1866): 382; *EMJ* 24 (September 15, 1877): 201; *EMJ* 27 (January 11, 1879): 30; *EMJ* 46 (November 24, 1888): 441; Ernest LeNeve Foster, "The Colorado Central Lode, a Paradox of the Mining Law," *Proceedings of the Colorado Scientific Society* 7 (1901–1902): 41–43.

Cincinnati lawyer-investor David Hyman figured in a case, testing the validity of the apex law, that became one of the most spectacular and expensive legal battles in Colorado history. The trouble arose in 1884 after the Emma and Aspen claims, which "sidelined" Hyman's Durant mine on Aspen Mountain, produced enormous quantities of rich ore (the Aspen mine reportedly taking out $600,000 in thirty days). Although Hyman's property had yet to hit the great bonanza, he charged that the silver vein coursing through neighboring claims rightly belonged to him by virtue of the apex, supposedly located in the Durant. Early in 1885 began a series of suits and countersuits that pitted Hyman and Charles A. Hallam of the Durant against Jerome B. Wheeler of New York and the other owners of the Emma and Aspen claims in litigation that would last three years, cost hundreds of thousands of dollars, and attract the finest legal and technical experts in the region to argue one side or the other. Hyman's forces enjoyed the legal acumen of Charles J. Hughes, Jr., and U.S. Senator Henry M. Teller, while Wheeler hired Thomas M. Patterson, the brilliant Charles S. Thomas, and J. M. Downing, each one a seasoned performer in mineral law cases.[33]

The reminiscences of David Hyman reveal the trouble, expense, anxiety, and tireless efforts behind the law suits involving the Durant claims. To raise money for the litigations, he almost immediately traded one-half of his interest in the property to Albert E. Reynolds, a Denver mine operator, for $25,000 to be used at once in gathering evidence and retaining Charles Hughes; then he extended an option on his mine to a trio of easterners for $360,000 more; and, finally, in 1887, he subdivided his interests still further by accepting $100,000 from Nathaniel K. Fairbank of Chicago, a millionaire manufacturer of soap and lard. As Hyman remembered it, he also became deeply concerned about the fairness of the upcoming trial in the federal district court and, though holding the "highest opinion" of Judge Moses Hallett who would preside, he had heard disquieting stories about jurors, sheriffs, and even judges being freely bribed in mining cases.

---

[33] The principal engineers in the cases were Wolcott E. Newberry for the apex and David W. Brunton for the sideline (Rickard, ed., *Interviews with Mining Engineers*, pp. 75–76). *MSP* 51 (October 17, 1885): 258; *EMJ* 44 (October 22, 1887): 299; New York Times, December 24, 1886, p. 1; Frank L. Wentworth, *Aspen on the Roaring Fork*, pp. 248–54; Len Shoemaker, *Pioneers of the Roaring Fork*, p. 139.

Later stating that he wished only to guarantee an "impartial" hearing for the Durant, he cleverly arranged for letters from Justice Stanley Matthews of the United States Supreme Court and Governor J. B. Foraker of Ohio vouching for Hyman's good character and had these documents personally delivered to Hallett by Senator Teller.[34]

Because of the mineral wealth involved, the crucial *Durant* v. *Aspen* case was described as the "most important mining litigation ever tried in Colorado." Before it commenced in December of 1886, Hyman and his chief counsel, Hughes, worked every day and night for seven weeks preparing their briefs, figuring that they had to make a "case as strong as that which is required in a criminal prosecution for murder." When rumors reached them that the sideliners would soon be defended in the pages of the *Denver Republican*, they speedily sent a representative to publisher Nathaniel P. Hill, reminding him that contempt of court might result, and Hill, good-naturedly taking the hint, blocked the offensive article. All of the strategy sessions and elaborate preparations paid off, for, in the several suits fought between 1885 and 1889, the Durant emerged the winner. One particularly successful tactic had been to discredit the witnesses for the sideline by showing their previous acceptance of the apex law and casting doubt on their veracity. When one witness for the Wheeler group testified that he had deliberately falsified a mining report to deceive eastern investors, since "he was not in this country for his health, and that such things had to be done in the mining business," Judge Hallett instructed the jury that any man so "destitute of principle" must be doubted on everything he said. Victor at Aspen and veteran of many trials as an attorney in Ohio, Hyman recalled twenty-nine years later that this fight had thoroughly fatigued him, adding that "no litigation equals a mining litigation in its intensity and bitterness."[35]

The litigation at Aspen might have gone on after 1887 had not

[34] Hallett was an exceptionally able jurist (see John D. W. Guice, *The Rocky Mountain Bench: The Territorial Supreme Courts of Colorado, Montana, and Wyoming, 1861–1890*, pp. 96–110). David M. Hyman, "The Romance of a Mining Venture," 1916 (typescript memoir lent to the author by the late Donald M. Hyman, New York), pp. 57–61.

[35] To prevent jury tampering, Judge Hallett brought to Colorado the chief of the United States Secret Service (Hyman, "Mining Ventures," pp. 47–58). *Hyman* v. *Wheeler et al.*, 29 *Federal Reporter*, 351 (1887); *MSP* 54 (January 1, 1887): 2.

the adversaries finally agreed to compromise their differences and spare the further loss of time and money in the courts. Even before the battle started in 1885, Hyman sent an agent to Jerome Wheeler in New York with instructions to sell the Durant for $70,000, but Wheeler, "badly advised" and "surrounded by unscrupulous men," according to his foe, refused the deal, thinking he could achieve his ends in court for a lot less. Nonetheless, by the end of 1887, both sides welcomed compromise; Wheeler had lost the legal fight and spent vast sums of money, while Hyman was weary of the protracted conflict, feared Wheeler would permanently close his Aspen smelter, and believed that further hostilities might delay construction of the Colorado Midland Railroad, another Wheeler enterprise. A compromise mining company resulted in 1888 and enabled gainful work to resume at the mines, with the Aspen property alone producing about $3 million in dividends over the next few years. Tallying up the costs of litigation, Hyman estimated that $600,000 had been spent, and this the Durant would have to repay before declaring any legitimate profits. From then on, thought Hyman, he could compromise every disagreement for no more than the cost of legal proceedings, and better still, he would "know exactly what the result would be without incurring the uncertainty of litigation."[36]

Many more mining financiers in Colorado reached the conclusion that compromise provided an easier and cheaper way out of conflict than the tortuously slow and complex procedures of the courts. Among the reluctant converts was William W. Glenn, an aristocratic Baltimore publisher and founder of the Maryland Jockey Club, who moved to Georgetown in 1872 after acquiring the Coldstream mine on Sherman Mountain for $500,000. Glenn was described by a reporter of the *New York Tribune*:

thousands of miles away from his relatives and friends, and from the comforts and luxuries which had become a second nature to him, this southern gentleman, in the decline of life, with money, friends, position,

[36] About the time that Hyman offered to sell part of the Durant to Wheeler, James J. Hagerman counseled Wheeler against the purchase: "Perhaps we can 'bust' their whole claim. The lawyers seem to think so" (letter dated June 10, 1885, James J. Hagerman Papers, Western History Research Center, University of Wyoming, Laramie). Hyman, "Mining Venture," pp. 61–67; *EMJ* 44 (October 22, 1887): 299; *EMJ* 45 (May 5, 1888): 329; *Aspen: Her Mines and Mineral Resources*, pp. 8–9.

popularity and large acquaintanceship, had left all to endure the hardships, the risks, the disappointments and the annoyances of mining.

Most of his hardships, Glenn told relatives, originated with the "lawless and godless" owners of the neighboring Phoenix mine who were trying to seize his property by suits over his patent at the same time that their "desperadoes" shot at his employees. Meeting this double-barreled challenge head-on, he retained the able mining lawyer Hugh Butler of Central City to handle the legal matters; hired Ross Raymond, who confirmed the richness of Glenn's vein, "whereas the others [veins] are all comparatively barren or insignificant, such as can be found in abundance on the side of Sherman Mountain"; and armed his miners and barricaded the mine to do battle with these "highwaymen," while Glenn himself traded blows with one of his enemies in a "rough and tumble fight" on a Georgetown street corner. In rapid succession, a worker on the Coldstream was murdered, boulders and dirt were dumped down a mine shaft, and the fifty-year-old Baltimore gentleman retaliated by smoking out the Phoenix men and hiring three deputy sheriffs to work as miners.[37]

His personal and financial resources strained to the limit after more than a year of conflict inside and outside the courts, Glenn agreed to compromise, deciding that his "life was not made for struggle" or for the "days and nights of harassing anxiety." Outraged and embittered by his experiences, he started to buy "*my own property* over again in order to get relief," purchasing a piece of disputed ground for $1,000 in 1874, and by mid-1875 declaring somewhat prematurely that nearly all the claims against him had been dismissed "without a trial." Yet at the time of his death in 1876 there remained thirty-six suits against the Coldstream, and an armed gang from the Phoenix immediately jumped the property. Ironically, the following year Glenn's brother found the title to the Coldstream tucked away in

[37] Rossiter W. Raymond, Report on the Coldstream mine [1873], Adelberg and Raymond Papers, Manuscript Division, New York Public Library; *EMJ* 14 (July 9, 1872): 27. Undated clippings from the *Georgetown* (Colo.) *Courier* and the *New York Tribune*; W. W. Glenn to John Glenn, November 23, 1873; Glenn to "my dear little boy," December 3, 1873; George H. Barrett to W. W. Glenn, January 12, 1874; Charles P. Baldwin to W. W. Glenn, July 21, 1874; all in Box 5, William W. Glenn Papers, Maryland Historical Society, Baltimore.

a land office at Central City and it showed no conflict; all suits were dropped, and the two properties united into the Consolidated Coldstream Mining Company.[38]

The principal effect of litigation—whether springing from honest disagreements over title or the apex or from those Raymond called the "litigating classes," who used the law for expropriation—was to impede the sound development of many mines and sap the financial strength of a capital-shy industry. In newer camps, particularly, legal disputes multiplied as young, greedy companies jostled and quarreled over the richest claims, frequently endangering their own survival but also explaining why lawyers flocked to every new mining town and rapidly set up a flourishing practice. Reflecting this characteristic, the *Engineering and Mining Journal* in 1878 predicted that a large part of Leadville's early profits would go to the lawyers, while a Cripple Creek promoter in 1895 gleefully invited eastern capitalists to his district now that the "retarding period" of litigation had passed. Typically, companies faced with legal problems hurriedly stripped the richest ore from their mines to furnish the "sinews of war" for lawyers and geological experts and likewise gouged out as much ore as possible in the event they had to surrender the property by court order. Often, capital earmarked for development of the mine was shifted into the legal war chest or, to avoid a suit altogether, went for purchasing surrounding claims, some worthless and many more carrying inflated price tags. With twenty-six suits pending against the Portland mine at Cripple Creek in 1895, the company decided to take the "bread from the lawyers" and bought seventeen neighboring claims for $1,025,000. And when a court fight became the choice, the costs could be equally staggering. The Iron silver mine at Leadville, for instance, once entangled in a web of sixty suits, spent over $500,000 trying to win the favor of the courts between 1878 and 1888.[39]

---

[38] So thoroughly disgusted did Glenn become with the Coloradans he met that he told his young son that there were no nice boys in Georgetown, only "little scamps," who were "all dirty, foul mouthed and godless" (to Willy, March 22, 1874, Box 5, Glenn Papers). Also, W. W. Glenn to Ann, January 20, 1874; W. W. Glenn to mother, November 17, 1875; W. W. Glenn to Willy, June 11, 1875; Petition of John Glenn to Baltimore Orphans Court, undated; all in Box 5, Glenn Papers. John Glenn to Will Glenn, June 25, 1879, Box 9, Glenn Papers.

[39] *EMJ* 24 (September 15, 1877): 201; *EMJ* 26 (December 7, 1878): 399;

In the opinion of many thoughtful observers of western mining in the last quarter of the nineteenth century, the industry would never attract its full share of capital or realize its potential for growth until cooperation replaced litigation and until the mining law of 1872 began to protect the miner and not succor the lawyer. The apex law was a disruptive element, said Raymond, that frequently worked "substantial hardship or even injustice" and caused the industry immeasurable waste and trouble. Some doggerel penned by the engineer in 1899 captured the drama of western trials in the mineral law and alerted absentee capitalists to the snares and expenses that might lie ahead:

> Sure, many such things may combine
> To make your mine not yours, but mine.
> If you don't buy me, fear the worst!
>
> .   .   .   .
>
> Thus was begun the famous case
> That filled the journals of the place,
> And thither called a mighty host
> From all the wide Pacific coast—
> A dozen lawyers on the side,
> And eminent experts multiplied;
> Maps of the biggest and the best,
> And models till you couldn't rest;
> Samples of rock and vein formation,
> And assays showing "mineralization,"
> And theories of that or this,
> And revelations of "genesis,"
> And summings-up of sound and fury.
> No matter now which party lost—
> It took the mine to pay the cost;
> And all the famous fight who saw
> Beheld, with mingled pride and awe,
> What science breeds when crossed with law.[40]

Rocky Mountain News, April 23, 1909; Walter R. Crane, *Gold and Silver*, p. 18; Henry R. Wray, comp., *Some Facts About Cripple Creek, Colorado: America's Greatest Gold Camp*, p. 10.

[40] Raymond read his poem "Science and Law" at a banquet of the American Institute of Mining Engineers, San Francisco, September 27, 1899 (reprinted in *MSP* 80 [April 14, 1900]: 400). Raymond, "Law of the Apex," 410–11, 442–44; Thomas A. Rickard, *The Romance of Mining*, p. 221.

Whatever perils threatened a mining enterprise, from financial finagling in the East to legal assaults in Colorado, the investor ultimately ran the risks and paid the bills. Concluding in 1877 that these pitfalls were often deeper and more dangerous than the mines, the *Engineering and Mining Journal* was amazed that the investor emerged "as an owner of anything except the clothes on his back."[41]

41 *EMJ* 24 (September 22, 1877): 219.

6

Colorado Offspring

THE eastern and foreign capitalist seldom worked alone in tapping the vast mineral riches of Colorado at the end of the century. Alongside him, even several steps ahead at times, strode the homegrown entrepreneur, marching to the same martial music of economic progress and sharing the same vision of imperial wealth, who recognized unrivaled opportunities to subdue the frontier, establish permanent communities, and make money by cooperating with outside mining investors. With the possible exception of Horace A. W. Tabor and Winfield Scott Stratton, who amassed regal fortunes from the mines in Leadville and Cripple Creek respectively and enjoyed a moment of national notoriety, most of these local operators remained obscure figures, constructing solid financial foundations slowly and unostentatiously while exercising considerable influence at the regional level. Whether railroad presidents, laboring men, or small-town merchants and lawyers, this diverse group of businessmen seized the opportunities arising from a vigorous economy based on large deposits of gold and silver metal and the influx of developmental capital from thousands of absentee investors. Inevitably this process enabled Coloradans to accumulate limited quantities of investment capital of their own, to acquire essential knowledge and experience in business methods, and to take the initiative in enterprises close to home.

As of 1866, a correspondent for the *New York Times*, appalled by the raw and litter-strewn mining camp at Central City, could write that "people come here only to make money not to live." Of course, he was right, to a great extent. Especially in the early years, Colorado's population fluctuated with the bright hopes and gloomy disappointments of immigrants who expected to stay in the territory only long enough to strike it rich before returning to their eastern homes. But the long-term development of the deep mines, reliant on eastern funds, fostered the growth of a stable population, which increased in leaps and bounds from an estimated 25,000 in 1861 to nearly 540,000 by 1900. Once their dreams of recovering a fortune from some hidden canyon in the West had evaporated, many of the newcomers to Colorado chose to stay nonetheless and work someone else's discovery or provide the mines with everything from dynamite to beef cattle, transforming the crude camps into bustling little towns that offered, to people with energy and ambition, employment and opportunities in banking, transportation, merchandising, and real estate. The mines alone needed scores of skilled and unskilled workers, and, when supplemented by millers, engineers, lawyers, and assayers, they formed the nucleus of a permanent population in these new industrial cities on the Rocky Mountain frontier.[1]

With a little luck and much hard work, some local men had been able to extend and develop their mines through the self-regenerating cycle of removing ore from the ground, selling it, and investing the proceeds in their properties. Eben Smith, a veteran of the California boom of the 1850s, arrived in Gilpin County in 1860, where he joined another pioneer, Jerome B. Chaffee, in working the Bobtail lode and running a small, primitive stamp mill. From the profits put aside in those operations, they loaned a neighbor $600 in 1862–1863 and acquired the right to mine his property for six months, during which time they recovered $35,000 and the means to purchase additional claims and begin new projects. Likewise, Canadian-born John F. Campion systematically reinvested his savings in newly discovered properties around Leadville, a practice which had brought him to Colorado from Nevada in the late 1870s, and subsequently he financed the opening of a great belt of gold ore that ran through his Ibex mine, giving the silver town a new lease on life in the 1890s. Though Campion's success never seemed assured, he shrewdly managed his accounts to keep a balance between the returns on one group

[1] The urban character of the mining frontier in the Rockies is described by Duane A. Smith, *Rocky Mountain Mining Camps*, pp. 42–77. "Croquis," *New York Times*, February 25, 1866, p. 2.

of mines and the expenditures on another, a feat that eventually paid
off handsomely in the Ibex.[2]

However, there were definite limitations on how far development work could go when the principal source of working capital
came from ore receipts, and the temporary disappearance of the ore
vein—"in cap" said the miners—the need for heavier machinery, or
indeed any grave complication could force the sale or abandonment
of a property. Irving Howbert, an early settler in Colorado Springs,
sold a large share in an exceptionally rich silver deposit on Fryer
Hill, Leadville, to a New York syndicate after a number of expensive
law suits bedeviled his Robert E. Lee mine. Even though the property
had earned $500,000 over a three-month period in 1879, Howbert
chose to sell and, among other things, to help in the construction of
an ornate opera house for Colorado Springs. At other times it may
have been simply the wisdom born of years in the mountains that induced local operators to sell their properties, including good producers, knowing that most mines had relatively short life spans. John
L. Routt, Colorado's last territorial and first state governor, bought
the Morning Star claim at Leadville for $10,000 at the end of 1877,
spending much of his time and money developing it, even taking
leave from his official duties in Denver to dig in the tunnels. In time,
he struck a blanket deposit of rich silver ore that reportedly paid him
$50,000 monthly. Despite this, in 1881 Routt sold the Morning Star
to easterners for $1 million and personally received over $600,000, a
small fortune that he increased through later investments in the state.
Both Routt and Howbert had acquired mineral properties at relatively
low cost, exploited them, and eventually sold out, preferring to take a
certain profit over the risk of future losses.[3]

Nor did Coloradans hesitate to part with their properties during

[2] Frank Fossett, *Colorado: A Historical, Descriptive and Statistical Work on
the Rocky Mountain Gold and Silver Mining Regions*, pp. 223–27; *Denver Times*,
November 6, 1902; William B. Vickers, *History of the City of Denver, Arapahoe
County, and Colorado*, pp. 361–63; Frank Hall, *History of the State of Colorado*,
4: 494–95; Carlyle C. Davis, *Olden Times in Colorado*, p. 196.

[3] Albert B. Sanford, "John L. Routt, First State Governor of Colorado,"
Colorado Magazine 3 (August, 1926): 83–84; William M. Thayer, *Marvels of
the New West: A Vivid Portrayal of the Unparalleled Marvels in the Vast Wonder-
land West of the Missouri River*, pp. 526–28; *EMJ* 25 (April 20, 1878): 276;
EMJ 32 (September 3, 1881): 156; Irving Howbert, *Memories of a Lifetime in the
Pike's Peak Region*, pp. 254–55.

the periodic mining fevers that swept the East, causing prices to sky-rocket and greenhorn investors to ascend the Rockies searching for a sure thing. Smith and Chaffee, who had succeeded in taking out large amounts of good ore from the mines around Central City, wasted little time selling their interests to New Yorkers at the height of the boom in 1864 and pocketed $50,000 apiece. Fifteen years later Horace Tabor eagerly, and probably fraudulently, took advantage of eastern excitement over the Little Pittsburg to dump his shares in the company shortly before it collapsed. Likewise, Winfield S. Stratton, the prospector made millionaire by the superb Independence mine at Cripple Creek, could not resist an offer of $10 million from a British syndicate in 1899. These few examples indicate that there were always mine owners in the West eagerly awaiting the approach of eastern financiers who, as some thought, quickly began "scattering greenbacks" in the territory like "an old uncle, just returned from India, who comes in with no end of rupees and a 'Bless you my children!' " The fact that wealthy outside capitalists invaded the West, buying up what they believed to be the best properties, eventually led to anger and frustration in some Colorado people, but it ought to be remembered that those who sold the mines speculated as much as those who bought, and from the viewpoint of a local operator, a mine placed in eastern hands meant quick profits for him and new development funds for his state.[4]

The inability of many absentee owners to conduct their business wisely and profitably often benefited local men who happily picked up the pieces of a moribund mining enterprise and, by avoiding the earlier mistakes, put the operation on a paying basis. The mining journals of the era bulge with examples of financial miscalculations in the East and managerial problems in the Rockies that drove companies into bankruptcy and enabled Coloradans to purchase mining properties cheaply, sometimes after costly improvements had been made. When several companies working on the Burroughs lode in Gilpin County closed down in 1869 and 1870 because of wanton expenditures and poor management, the superintendent of the Ophir

[4] *AJM* 7 (April 24, 1869, and May 29, 1869): 264, 340; G. Thomas Ingham, *Digging Gold Among the Rockies*, p. 402; Fossett, *Historical, Descriptive and Statistical Work*, p. 227; Clark C. Spence, *British Investments and the American Mining Frontier, 1860–1901*, pp. 28–29.

mine, Colonel George E. Randolph, united the best properties under his able direction and earned steady returns on the group for a number of years. In 1872 Randolph's financial successes helped carry him into the mayor's office in Central City, although he later shifted his business talents to Denver, where he took a leading role in expanding and modernizing the Denver City Street Railway. John and Charles Briggs, also residents of Central City, sold their mine to New Yorkers in 1864 and watched as the new owners poured more than $50,000 into developing and improving the property, only to abandon it four years later as a result of the high costs. Seeing a fresh opportunity, the Briggs brothers repurchased the mine at a sheriff's sale (for much less than they sold it) and quickly had it producing again, helping them to purchase neighboring mines and add to a growing list of investments. In a like manner, John Quincy Adams Rollins, one of the pioneers in the territory, earned a reputation for being the "most energetic man in the whole world" by relocating and buying abandoned claims in the Gold Dirt district of Gilpin County where eastern efforts had failed. Over the years of the 1860s and 1870s, he gained control of nearly the whole productive area, extending to 20,000 feet on several lodes, built a town, and incorporated a mining company with a capital stock of $5 million, which he then promoted in the East.[5]

After a leisurely sojourn in the Rockies near the end of the mining boom of the 1860s, Massachusetts newspaperman Samuel Bowles noted that "reliance upon eastern capital, was, if not at its height, still reigning but, with signs of decay and threatening despair." There was no reason to be demoralized, however, for in his opinion these would also be the "kernel years" for Colorado, when eastern failures challenged local people to come forward, secure the mines, and run them conservatively for the benefit of the home economy. To a greater or lesser extent, varying with the time and place, this happened in Colorado throughout the first forty years of the industry, and idle or abandoned mines were revived by men on the scene, with or without

[5] Fossett, *Historical, Descriptive and Statistical Work*, pp. 215–18; Ovando J. Hollister, *The Mines of Colorado*, p. 222; *AJM* 7 (March 6, 1869): 147; Frank Fossett, *Colorado, Its Gold and Silver Mines, Farms and Stock Ranges, and Health and Pleasure Resorts*, pp. 344–46; *Rocky Mountain News*, October 23, 1911; *MSP* 35 (September 22, 1877): 178; John Q. A. Rollins, Jr., "John Q. A. Rollins, Colorado Builder," *Colorado Magazine* 16 (May, 1939): 112–14.

the cooperation of eastern owners. Recognizing a rare opportunity in 1867, a Boulder miner urged his neighbors to transfer heavy machinery and mills left behind by eastern companies to places where they could do more good—quite obviously, to mines being run by residents. In the mid-1880s new stories from Leadville reported "wholesale" machine stealing and unauthorized removals of expensive equipment and tools; one owner claimed that he had lost three large steam-powered hoisting engines in two years, while others said they returned after work stoppages to find "every movable piece missing," including entire shaft houses. Yet these remained isolated incidents, and thefts of this magnitude were never generally accepted or condoned by miners as a way of gaining a greater share in the mining business of their region.[6]

Of greater consequence as a means of bringing together resident and nonresident mining men was the system of mine leasing, which appeared very early in Colorado and with the passage of time evolved into one of the principal methods for exploiting the gold and silver deposits. One journal estimated in 1898 that fully three-quarters of the productive mines at Cripple Creek were in the hands of lessees. Fundamentally, the lease was a contract between the mine owner and one or several persons to work all or a piece of his property over a stated period of time, with the lessees receiving a percentage of the returns on the ore removed and delivered to the mills. Just how the system originated is somewhat unclear, but O. J. Hollister reported that as early as 1860 one-half of the famous Gregory discovery was leased to a group of miners calling themselves the American Mining Company, beginning a series of leases on the property and leading to its adoption elsewhere in the district. Charles W. Henderson in his generally reliable history of Colorado mining took this to mean that leasing was "introduced by Americans," yet the system closely resembled the "tribute work" done in Cornwall, England, where laborers often kept a part of the ore they dug, and hundreds of Cornishmen helped to spread this practice in the American West. Considering the similarity between leasing and tributing, many Cousin Jacks made the transition easily and successfully, as did a group of Cornishmen

6 Samuel Bowles, *Our New West: Records of Travel Between the Mississippi River and the Pacific Ocean*, pp. 79–80; *AJM* 3 (May 4, 1867): 107; *MSP* 49 (December 13, 1884): 375.

who leased a Central City mine in 1873 and in less than two years earned over $100,000 for their return trip to Cornwall in style. The question of origins aside, it became clear that Colorado adopted leasing on a large scale, perhaps surpassing the other mining regions in this regard, and through expansion and modification made the practice an integral part of the industry, with advantages and disadvantages for owners and lessees alike.[7]

Leasing took root in Colorado gradually, spreading from district to district as local conditions changed. Typically, it made the greatest gains in camps nearing old age, at least in terms of the quantity and quality of the ore, and when a great deal of unprofitable dead work was required to reach the seams of metal, or when companies reacted to an economic setback brought on by higher costs, capital shortages, or waste and inefficiency on the part of management. The lease system arrived in Clear Creek County in 1882, said one observer, after many eastern companies had "touched the last dollar" through the "incapacity or inability of mine owners to make both ends meet by running the mines themselves," and because the mines were simply getting poorer and harder to work. Referring to a property in Summit County, a Denver newspaper noted that "leasers" (a western colloquialism for lessees) take over after a "stock jobbing deal" and "enrich themselves where a corporation could find no ore." Cripple Creek, though still a rich and promising field that had yet to reach its peak, adopted the system in the mid-1890s as a result of the financial panic of 1893 and labor troubles in 1894, and, according to the Denver Republican, it proved to be the "salvation" of the district and demonstrated to capitalists that the camp merited further investment.[8]

Companies that chose to negotiate a lease rather than employ a force of wage workers did so in the belief that leasing could save

[7] Hollister, Mines of Colorado, p. 64; Charles W. Henderson, Mining in Colorado: A History of Discovery, Development and Production, U.S. Geological Survey Professional Paper 138, p. 27; EMJ 55 (March 4, 1893): 193–94; Fossett, Historical, Descriptive and Statistical Work, p. 280; MSP 49 (September 27, 1884): 196; MSP 79 (September 16, 1899): 310.

[8] Leasing increased at Leadville, Creede, Aspen, and other silver districts in Colorado after the collapse of the silver market in 1893 (see Denver Tribune quoted in MSP 45 [July 29, 1882]: 65; Denver Republican quoted in MSP 71 [September 14, 1895]: 171). Hall, History, 4: 411; N. E. Guyot, "Cripple Creek: An Inside Story," Engineering and Mining Journal-Press 118 (December 20, 1924): 966; D. Bauman, King Carbonate, Leadville, Colorado, p. 5.

them money and increase their profits. For instance, while maintaining possession of the property—a company superintendent or agent usually remained at the mine—the owners could shift a large proportion of the risk and expense from themselves to the lessees, who contracted to do a stated amount of development work, keep a certain number of men on the job, and pay as much as 40 to 50 percent royalty if high-grade ore was removed. From the mill returns on the ore, the owners deducted their royalty, as well as any charge for the use of tunnels and hoisting machines, tools supplied, and the cost of assaying. Beyond these technical aspects of a lease contract, company officials generally believed that persons working for themselves were more likely to labor longer and more efficiently, produce greater quantities of ore, use supplies more economically, and take more chances to find pay ore than paid hands, who "proverbially understood" the ways of "shirking" work in a mine shaft. President Rossiter Raymond of the Chrysolite mine decided to use the leasing system at Leadville in the early 1890s so that the difficult task of exploration was carried on by somebody other than his own financially fragile company. With his characteristic bluntness, Raymond, a long and bitter opponent of organized labor, added another reason. "The exorbitant wages demanded by the miners' union in the west," he argued "bear with special severity upon the execution of all dead-work, which is not immediately productive, and upon all mining, the profit of which is at all doubtful. By voluntary leases and contracts this tyranny is evaded."[9]

Not to be overlooked, however, were the many opportunities the system offered to a lessee. Once a mine was opened and equipped, its capital needs were greatly reduced, and if, as frequently happened, the lease went to men who had worked on the property previously,

[9] The *EMJ* placed the average royalty for Colorado leases at 20 to 25 percent on high- or medium-grade ores and 10 percent on low grade (53 [April 16, 1892]: 421). The *Leadville Herald-Democrat* noted that the loudest objections to leasing came from "petty" officials at the mines who did little or nothing for the huge salaries paid them by absentee companies (*MSP* 52 [April 24, 1886]: 275). Benjamin B. Lawrence, "Notes on the Lease or Tribute System of Mining, as Practiced in Colorado," *Transactions of the American Institute of Mining Engineers* 21 (1892–1893): 912–13, and comments by Rossiter W. Raymond, 916–18; Francis T. Freeland, "Mining Leases," *Transactions of the American Institute of Mining Engineers* 25 (1895): 112; Thomas A. Rickard in Richard P. Rothwell, ed., *The Mineral Industry, Its Statistics and Trade, In the United States and Other Countries to the End of 1894*, p. 644.

they often knew more about the location of pay ore than the owners or managers, thus avoiding much of the time-consuming and unproductive deadwork. For these reasons alone, local miners were usually anxious to take leases on "exhausted" mines, especially when that meant mismanaged ones, knowing that a good possibility existed to "clean up a reasonable fortune from abandoned drifts and stopes." Owners of the Lucky Guss mine at Cripple Creek worked their property for two years, and losses, due mostly to extravagance, mounted rapidly. Seventeen miners took it over and in a few months had the mine producing well and paying good returns—a "striking illustration," said the *Mining and Scientific Press* in 1898, of "success and failure resulting from two different methods of mine management." Moreover, even though lease contracts set down strict legal obligations, absentee companies were hard pressed to enforce them because of the distance, and lessees were known to take ore from unleased sections of a property, pilfer high-grade ore, ignore requirements for the proper development of a mine and general maintenance, and cheat on many of the provisions in the contract, sometimes with the connivance of local ore buyers. For instance, Norvin Green "could not explain how I ever came to sign" a contract that permitted lessees to remove ore from anywhere on his property, work twice as long as the agreement provided, recover all of their expenses, make $10,000, and in addition, acquire a block of shares in the United Claims Mining Company, of which Green was president. On other occasions, particularly toward the end of a contract period, lessees "gophered" the mine by picking out the best ore at the same time that they reduced their costs to a minimum by shutting off water pumps, cutting back on the timber used for shoring up the tunnels, or letting the roads leading to the mine become so pot-holed that shipments out had to be suspended. T. A. Rickard reported from Cripple Creek in 1896 that lessees usually threw a "property back into the hands of the owner in a depleted and dilapidated condition."[10]

The expansion of mine leasing, notably in the 1890s, afforded

[10] *MSP* 70 (February 2, 1895): 65; *MSP* 76 (June 11, 1898, and June 18, 1898): 623, 638–39; Rickard quoted in *MSP* 72 (April 11, 1896): 285; *EMJ* 60 (October 5, 1895): 322; Green to Charles H. Emerson, January 16, 1889, January 23, 1889, January 29, 1889, Letterbook 4, Norvin E. Green Papers, The Filson Club, Louisville, Ky.

new opportunities for residents who had accumulated a modest amount of capital to push the system beyond small groups of mine laborers working for themselves and responsible directly to the owners or their western agents. It became obvious that in some of the larger and longer leases the need for working capital to pay laborers, buy tools and supplies, or purchase a costly piece of machinery seriously strained the personal resources of the lessees and threatened their ability to fulfill the contract. At such times, local businessmen of every description frequently stepped forward to help with these financial problems, agreeing to share in the risks and profits of the lease, out of which emerged a new business alliance between experienced, practical miners who directed the underground work and town merchants or others who handled the financial end of the project. A local paper noted in 1886 that Leadville miners had been joined by about four-fifths of the town's merchants and professional men in leasing the better old mines and even opening a few new properties; the pick and shovel miners actually operated the lease, while their businessmen partners supplied equipment and wages and kept the account books. According to that newspaper, leasing at Leadville had developed into a lucrative outlet for the surplus capital of townspeople at the same time that it gave a lift to the area's sagging economy.[11]

Nearly all of the major mining districts in Colorado produced one or more leasing companies organized and operated by local entrepreneurs. On the whole they were small business ventures limited to one mine for a specified period of time, although a few grew into potent enterprises by leasing large sections of a camp and bringing together several different properties under one management. Of these, among the most successful was the Union Mining and Leasing Company, formed at Leadville in the depression year of 1893 by Seeley W. Mudd, Robert B. Estey, and Franklin L. Ballou, all of whom lived in the city and had years of experience with the mines. Taking note of the trouble that companies on the east side of Fryer Hill encountered in litigation, lower grades of ore, rising water levels in the tunnels, and the waste and duplication that resulted from many small com-

---

[11] The *Denver Post* reported that "Pools" of miners and local businessmen in Gilpin County succeeded in reviving many old and abandoned mines in the mid-1890s (*MSP* 77 [August 13, 1898]: 162). *Leadville Herald-Democrat* quoted in *MSP* 52 (April 24, 1886): 275.

panies working independently, they leased and consolidated the properties (approximately 170 acres), and managed the whole as one operation, which in 1895 employed two hundred men. For three years, the company prospered, paying regular dividends on its $500,000 capitalization, and in the seventeen months before its disappearance in the Leadville strike of 1896, it produced $1,370,000 in silver ore. Mudd and Estey reinvested their earnings from the Union Company in other mine leases around Leadville and acquired additional leasing interests in the state. Eventually, Seeley Mudd moved to bigger things, earning the respect of the international mining community for developing the extraordinary copper deposits on the island of Cyprus in the eastern Mediterranean.[12]

There were times when local entrepreneurs negotiated leases in hopes of heading off the abandonment of mines in the area by absentees, discouraged perhaps by lower returns or labor problems, and with the intention of attracting new capital to the region by proving that the mines could still produce at a profit. The local citizenry, particularly businessmen with a stake in the town and its economic well-being, needed no reminder that their personal fortunes rested on the success or failure of the mines. After several eastern and Denver financiers, among them Eben Smith and David Moffat, refused to reopen the so-called downtown group of mines at Leadville, which had become flooded during the strike of 1896, a syndicate of local businessmen headed by mine manager A. V. Bohn and bank president Albert Sherwin decided to do the job themselves in 1898. With financial support from dozens of merchants and tradesmen, the Home Mining Company drained the properties at the foot of Carbonate Hill, sunk into rich ground, extended its tunnels beneath the streets and alleys of Leadville proper, and employed over one hundred men.

---

[12] Another Estey project at Leadville, the Coronado Leasing and Mining Company, reputedly attracted Admiral George Dewey and General Elwell S. Otis, military governor of the Philippines, as stockholders (*MSP* 79 [October 21, 1899]: 466). *MSP* 76 (January 29, 1898): 124; *MSP* 79 (December 16, 1899): 695. Alva Adams, a former governor of Colorado, and Otto Mears employed over two hundred men in a leasing operation at the Ute and Ulay mines, Lake City, in 1893 (*EMJ* 56 [November 25, 1893]: 551). *EMJ* 59 (April 13, 1895): 349; *EMJ* 62 (July 24, 1896): 85; Thomas A. Rickard, ed., *Interviews with Mining Engineers*, pp. 396–97. On Seely Mudd, see David Lavender, *The Story of Cyprus Mines Corporation*, pp. 3–11.

Doing far better than showing Leadville to be no ghost town, the Home Company raised its capital stock from $50,000 to $2 million as success followed success, and it paid $240,000 in dividends, making it one of the most sought after Colorado mining stocks by the turn of the century. On a smaller scale, home companies also sprang up in the deflated silver mining towns of Aspen, Creede, and Idaho Springs between 1899 and 1902 as their residents tried to revive the mines and restore the prosperity of an earlier day.[13]

It would be wrong to conclude that Coloradans barely scratched the surface of their own mineral wealth before selling out to absentees or that they merely scavenged the deep mines after outside companies had left. In fact, throughout the early decades of the industry, there were vigorous attempts by groups of local men to finance mining ventures. Between 1867 and 1868 most of the capital employed on several silver deposits near Georgetown came from the black residents of that town and elsewhere in the territory who had organized the Red, White and Blue Mining and Reducing Company under the leadership of a former slave from the Missouri lead mines. The Marshall Silver Mining Company successfully placed nearly all of its capital stock in Georgetown, Central City, and Denver for the purpose of constructing a deep tunnel through silver-rich Leavenworth Mountain at Georgetown. And in the 1870s, a "lively competition" reputedly began between local and eastern capitalists in buying mining ground and building mills around Central City. Yet none of these investments matched the financial ties that grew between Colorado Springs, the fashionable resort city at the foot of Pike's Peak, and Cripple Creek, the boisterous gold camp fifty miles to the west at the top of the Continental Divide. Many of the first discoveries made in the grassy hills around Cripple Creek were purchased by Colorado Springs businessmen when a slowdown in the city's growth rate freed some capital for investments other than in real estate. Idle real estate brokers, such as Charles L. Tutt, Spencer Penrose, and W. P. Bonbright, began new careers in mining promotion and finance as a re-

[13] *Portrait and Biographical Record of the State of Colorado*, pp. 251–52, 1247–48; Jay F. Manning, *Leadville, Lake County and the Gold Belt*, p. 99; *EMJ* 66 (October 22, 1898): 496; *EMJ* 73 (January 4, 1902, and May 17, 1902): 9, 706; *MSP* 77 (August 13, 1898): 151; *MSP* 79 (July 8, 1899, and October 7, 1899): 52, 410; *MSP* 84 (April 12, 1902): 209.

sult of wise purchases at Cripple Creek in 1891 and 1892. Succeeding years found that these financial links with the mining backcountry had increased; the *Engineering and Mining Journal* reported in 1895 that some eastern financiers grumbled about the large number of good mines controlled by Colorado Springs people, and the journal estimated that in just two weeks of that year $300,000 of local capital flowed into the mines. Over 75 percent of the mines and prospects were owned and managed from Colorado Springs, T. A. Rickard flatly stated in 1898. Moreover, by the end of 1895, four mining-stock exchanges, 275 mining brokers, and daily sales of over 1,000,000 shares attested to the active role townspeople were taking in the promotion and sale of stock to outsiders. Readily seeing a parallel in the fabulous gold fields of South Africa, the *New York Times* touted Colorado Springs as the "Johannesburg to the Cripple District."[14]

By 1900 the fuller dimensions of this financial bond had begun to show up in the number of Colorado Springs residents who had hit pay dirt in the hills at Cripple Creek. Besides the good fortune that had smiled on several poor prospectors, such as Winfield Stratton in the Independence and James Doyle and James Burns in the Portland —each a rags to riches tale—there were the small-town businessmen, relatively unknown and usually overshadowed by the more glamorous aspects of the mining frontier, whose keen sense for the right investment and talent for organization brought them power and prosperity. During the fifteen years prior to 1910, Frank, Harry, and Warren Woods, real estate brokers in Colorado Springs, fashioned a small empire at Cripple Creek, under the name of the Woods Investment Company, by consolidating many smaller mines and purchasing blocks of shares in several of the most productive enterprises. They engineered the largest consolidation at Cripple Creek as of 1902 when

[14] The historian of Colorado Springs, Marshall Sprague, found that only a few of Cripple Creek's richest mines were owned by nonresidents during the first twenty-five years of the camp (*Newport in the Rockies*, pp. 167–68). *EMJ* 8 (July 13, 1869): 22; *EMJ* 20 (December 18, 1875): 600–01; *EMJ* 60 (October 5, 1895): 329; *MSP* 16 (January 18, 1868): 37; Thomas A. Rickard in "Colorado: Resources and Attractions of the State," *Bankers' Magazine* 57 (July, 1898): 203–05; *New York Times*, December 2, 1895, pp. 4, 9; Guyot, "Cripple Creek," p. 937; William Weston, *The Cripple Creek Gold District . . . How It Was Formed (In Plain English)*, p. 8.

they brought together eight corporations into the United Gold Mines Company, which possessed 375 acres of mineral ground and was capitalized at $5 million, although this venture represented only a small portion of 1,500 acres and forty-two companies controlled by the brothers. James R. McKinnie and Albert E. Carlton, both of whom got their start in freighting supplies to the gold camp, became heavy investors in the mines and built a considerable fortune after 1896. McKinnie became a principal shareholder in the Moon-Anchor and Portland mines, two of the district's most valuable properties. Carlton in 1908 purchased the United Gold Mines, which then embraced one hundred mines, making him the second largest landholder in the area. A slightly different path had been followed by William Lennox, a resident of Colorado Springs since 1872, who at the time of the ore discoveries over the range was a prosperous coal dealer with a virtual monopoly. He reportedly bought the first claim made at Cripple Creek, the El Paso, in 1891, and thereafter became a leading mining financier. This evidence, along with the other indications of regional economic interest at Cripple Creek, no doubt explained why the *Mining and Scientific Press* concluded that of the $18 million in dividends paid by those mines in 1901, about $16 million went to Coloradans.[15]

The combinations of Colorado men that managed profitable leases or played a major role in opening the gold field at Cripple Creek reflected the regions' growing and maturing economy. Years of cooperation with absentee investors, of studying their financial methods for better or for worse while absorbing a part of the funds they expended in the state, had been the crucible within which local entrepreneurs were tempered, and it prepared them for taking greater advantage of the mineral wealth beneath their feet. Many frontier lawyers, at least those specializing in mining litigation where eastern corporation paid munificent fees, acquired valuable interests in the mines. Henry M. Teller, Central City lawyer and longterm U.S. Senator (1876–1882 and 1885–1909), invested in Gilpin, Clear Creek,

[15] Marshall Sprague, *Money Mountain*, pp. 69, 168–70; *EMJ* 66 (September 3, 1898, and December 4, 1898): 286, 766; *EMJ* 73 (May 24, 1902): 738; Mabel Barbee Lee, *Cripple Creek Days*, pp. 235–38; *MSP* 82 (January 19, 1901): 46; *MSP* 84 (February 22, 1902, and April 19, 1902): 107, 233; Harry J. Newton, *Yellow Gold of Cripple Creek*, pp. 39–41.

and Boulder counties during the 1860s and 1870s. Thomas M. Patterson, another illustrious member of the Colorado bar, admitted to losing over $70,000 in the mines of his state. General William J. Palmer, president of the narrow-gauge Denver and Rio Grande Railroad that wound through the mountains in pursuit of booming mining camps, joined other officials of the line in purchasing mining property at Ruby Camp and Silverton. Local capital of this kind was also generated by the expanding smelting industry, dependent, of course, upon the deep-level mines constructed with eastern money. Both the pioneer smelter Nathaniel P. Hill and his assistant Henry R. Wolcott possessed large mining interests all over the state, and August R. Meyer and Edwin Harrison, builders of the first smelter at Leadville in 1877, remained active investors there, returning to the mines some of the profits earned in ore production.[16]

Judging by the scope and success of their investments, surely the most formidable group of resident financiers was the combination of David H. Moffat, Jerome B. Chaffee, and Eben Smith. They cooperated on a number of mining and business ventures until Chaffee's death in March, 1886, after which Moffat and Smith often operated as a team. The imprint they left on Colorado reminds one of the sway held over Nevada's Comstock Lode by the famed Bonanza Kings, albeit with important differences: they never matched the latter's immense personal wealth, estimated at $50 million; the source of their fortune cannot be traced to a single mining enterprise; their investments invariably stayed in Colorado; and, regrettably, much less is known about them. Nonetheless, from the scanty information available, it appears that Moffat was the steady, hard-headed man of business, Chaffee was wily and energetic, and Smith was one of the most able practical miners and managers in the Rockies. Moffat's name, with one or another of his partners, is associated with nearly every major mining camp from Rico to Leadville, chronologically from the development of Central City to Cripple Creek.[17]

[16] Elmer Ellis, *Henry Moore Teller: Defender of the West*, p. 55; John F. Graff, *"Graybeard's" Colorado: or, Notes on the Centennial State*, pp. 81–82; *EMJ* 29 (April 10, 1880, and April 24, 1880): 260, 284; *EMJ* 41 (May 29, 1886): 396; *EMJ* 69 (May 26, 1900): 626; Fossett, *Gold and Silver Mines*, pp. 526–31; U.S. Congress, Senate, Committee on Mines and Mining, *Cost of Production of Gold and Silver Bullion*, 52nd Cong., 2nd Sess., 1893, p. 15.

[17] For the story of John W. Mackay, James G. Fair, James C. Flood, and

One key to understanding the success of these Coloradans as investors is found in the way they utilized the natural advantage of close proximity to the mines. In addition to being easily accessible to all sorts of impecunious promoters, the partners quickly learned about new mineral regions in the state that enabled them to buy into recent discoveries before the influx of eastern and foreign money drove prices up. Denver banker Moffat realized the value of bedrock investments and so employed a small legion of prospectors who, under the direction of Eben Smith and a few mining engineers, regularly fed him a stream of investment possibilities. Among these "gold hounds," as a contemporary labeled them, was Nicholas Creede, for much of his life a hard-luck prospector. His discovery of the Holy Moses mine in 1889 ignited a silver boom in the valley of the Rio Grande. At various times Creede had been grubstaked by Moffat, and it was to him that Creede sold a controlling interest in the Holy Moses, which Moffat and Smith then developed into a leading producer. Likewise, all three men apparently listened patiently to the never ending spiels of prospectors and promoters with claims to sell, sometimes with splendid results; Chaffee purchased his interest in the Little Pittsburg this way, and Moffat and Smith followed a similar course into the valuable Rico-Aspen mines on Dolores Mountain at Rico. Typical of the unsolicited contacts was the "liberty" taken by one promoter who offered Moffat a share in his property after learning that the Denver financier was always "desirous of helping young men" start a mining company. And a Boulder operator wrote Eben Smith secure in the knowledge that "yourself and partners are always looking for properties to open and make into mines."[18]

When conventional means failed to acquire good mineral

William S. O'Brien, the "big four" of the Nevada Comstock Lode, see Oscar Lewis, *Silver Kings.*

[18] John Campion reputedly bailed out indigent prospectors from Denver jails and probably acquired valuable information about new discoveries from these grateful friends (Amanda Ellis, *The Strange, Uncertain Years: An Informal Account of Life in Six Colorado Communities*, p. 200). Edgar C. McMechen, *The Moffat Tunnel of Colorado: An Epic of Empire*, 2: 75–76; Hall, *History*, 2: 435–36; James Pourtales, *Lessons Learned From Experience*, p. 158. James Cowie to Smith, August 16, 1894.; William M. Brown to Moffat, October 12, 1894; Alexander Gullett to Moffat, June 18, 1895; all in Box 1, Eben Smith Papers, Denver Public Library. Smith to Moffat, March 22, 1892, Box 2, Smith Papers.

ground, Moffat and Chaffee resorted to shrewd and unscrupulous tactics in their determination to buy a property, occasionally employing litigation or the threat of it as a weapon in the conflict. One observer credited Moffat with an "admired coup" at Cripple Creek for encouraging claim holders around the rich Anaconda property to bring an injunction against it based on prior location, whereupon he cunningly swooped in to purchase the hard-pressed mine at a huge discount. In this same period of the early 1890s, N. E. Guyot traded one of his Cripple Creek claims to Moffat for a block of shares, only to charge later that he had been badly "skinned" in the transaction. Critics condemned Chaffee as a "mine butcher" in 1876 and 1877 for the way in which he united several legal claims against the Caribou silver mines and relentlessly drove them through the courts until the Dutch owners capitulated and permitted him to buy the mine at a sheriff's sale. Approximately fifteen years before Chaffee faced the charges of angry shareholders that he had manipulated the stock of the Little Pittsburg, Henry Teller, during a heated political controversy in 1865, caustically nicknamed him "Joe Bagstock" Chaffee because of his aggressive techniques in the mining business—but also hinting that Chaffee may have plundered a few enterprises as well.[19]

David Moffat, for all of the actual and alleged skulduggery, displayed a genius for multiplying, integrating, and consolidating his interests at a mining camp so that each part depended on and supported the other, enabling him to cut costs and increase his profits. By purchasing small, contiguous claims and combining them under one management, he eliminated wasteful duplication and increased the efficiency of his operations, which often spelled the difference between profit and loss when handling leaner grades of ore. At Leadville, consolidations engineered by Moffat and Smith, particularly their extensive holdings on Carbonate Hill where they mined silver and lead, made them the most influential figures in the district by the early 1890s. Their importance in Leadville was magnified further by Moffat's role in constructing the Bi-Metallic Smelter in 1892 to treat the lower-grade ores which were increasingly becoming the bread-and-

[19] Pourtales, *Lessons Learned From Experience*, pp. 159–60; Duane A. Smith, *Silver Saga: The Story of Caribou, Colorado*, pp. 63–66; Guyot, "Cripple Creek," p. 936; *New York Times*, June 14, 1877, p. 2; Teller quoted in Elmer Ellis, *Teller*, p. 72.

butter of local mines. Yet it was at Cripple Creek that Moffat fashioned a fully integrated system of business ventures, including some of the richest mineral properties there, such as the Raven, Victor, and Anaconda mines, a large stake in the Florence and Cripple Creek Railroad, and a treatment plant near Florence that utilized the radically new cyanide process to leach gold from the ore. Thus, ore removed from his mines traveled over his railroad to his mill, giving him control over every major stage in the production of gold. Finally, as president of the mighty First National Bank of Denver from 1880 to 1911, Moffat had countless opportunities to draw upon the resources and reputation of that institution to promote his entrepreneurial activities. In 1900, for instance, the United States comptroller of the currency objected to the bank's habit of making heavy loans to mining companies, many of which proved to be the president's own, including the Victor and the Anaconda.[20]

The focal role Moffat played in the state's progress and prosperity earned him the praise of a variety of Colorado historians and commentators. One boasted of his full or partial ownership of more than one hundred mines, another lauded him as the "first civilian of the state," and a mining reporter simply thought that his name had become a "Colorado household word." Implicit in these tributes was an important fact: most of the money Moffat made in Colorado, beginning with his bookselling days in Denver, had been reinvested in the state, a point made crystal clear by Eben Smith in 1894 when he informed a New Mexico promoter that "we have refused to go outside Colorado as a rule to invest in the mines." Of equal moment, though generally glossed over by enthusiastic local boosters, was that nearly all of the Moffat-Smith mining operations were actively promoted outside the state and greatly relied upon absentee capital for development. In addition to the famed Little Pittsburg, there were also large sums of eastern money in their other mines at Leadville, as well as substantial amounts of French capital in the Victor at Cripple

---

[20] Moffat also skillfully integrated his mining and transportation interests at Creede in the 1890s (see Nolie Mumey, *Creede: History of a Colorado Silver Mining Town*, pp. 49–50). Copy of unsigned letter from the bank to the U.S. Comptroller of Currency, August 29, 1900; Moffat to Frank A. Gardener, October 13, 1897; Letterbook 1884–1902; all in David H. Moffat, Jr. Papers, First National Bank of Denver Archives. Hall, *History*, 3: 431–32; McMechen, *Moffat Tunnel*, 1: 80–82; *EMJ* 54 (August 13, 1892): 159.

Creek. Furthermore, the strength of Moffat's reputation as a resident financier sprang less from his mining than his railroad investments, which included the Denver and Rio Grande, the Denver Pacific, and the Denver and Salt Lake (the Moffat Road), or his exalted place among bankers in the state. In 1880, he reportedly owned 27,000 acres of Colorado land and $200,000 worth of Denver real estate, and, over the years, he acquired shares in the banks at Cripple Creek, Victor, Hot Sulphur Springs, Aspen, and Durango, owned one-half of the Denver Tramway System, and helped finance the Denver Union Water Company to supply the growing metropolis. Without question Moffat found the hardrock mines attractive and profitable, but, in view of the diversity of his investments, he never dealt in them exclusively nor did he hesitate to share their risks and expenses with outsiders. Along with scores of local businessmen, he shrewdly balanced his mining ventures with the steadier returns from banks, railroads, smelters, utilities, and real estate.[21]

From the evidence available, it seems that a substantial amount of the money generated by the mining industry in Colorado never returned to the mines in the form of local investments but was channeled into less volatile enterprises that promised to grow with the region and offered a firmer financial footing than the precious-metal mines. The tendency to hedge on the instability of the mines by putting their investment capital into a variety of business projects characterized the activities of a number of local entrepreneurs. Mining lawyer Henry M. Teller counterbalanced his silver ventures with toll road and railroad stocks in the early 1870s. Walter B. Devereaux, a gifted mining engineer and general manager of the Aspen Mining and Smelting Company from 1883 to 1893, invested in the coal mines of Garfield County, helped found the Roaring Fork Water, Light and Power Company, and was interested in banking and real estate in Glenwood Springs. And lastly, John F. Campion, the Leadville mining man, owned shares in Leadville and Denver banks as well as power companies in Colorado, Utah, and Missouri, acquired stock in local railroads, became the president of a Wyoming cattle company,

21 McMechan, *Moffat Tunnel*, 1: 65–82, 104–05; *New York Times*, March 19, 1911, p. 11; *EMJ* 55 (May 20, 1893): 469; *EMJ* 91 (March 25, 1911): 630; Vickers, *City of Denver*, pp. 509–11; Hall, *History*, 3: 128–29; Smith to John S. Crawford, September 27, 1894, Box 1, Smith Papers.

and worked to establish a valuable sugar-beet industry in Colorado. In light of these many nonmining investments made by residents, the *Silver Plume Silver Standard* stoutly defended absentee capitalists from critical Coloradans in 1888:

It is a fact worthy of remembrance that non-resident investors are the men who openly acknowledge their connection with mining, and this more than counterbalances all the mistakes they have made in the management of their properties here. It should also be borne in mind that Colorado businessmen have grown rich through the absorption of the money lost by nonresident investors, and hence there should be no grumbling at the few million that flow back to them each year through the few enterprises which are now steadily profitable to them. The bulk of money required by the purchase and operation of mining properties has gone into the pockets of the residents of Colorado, and yet our people fail to zealously work for the more rapid advancement of the mining industry of the state. They talk up real estate and manufacturing and cry down mining as an unsafe and risky business.[22]

The movement of Colorado capital into safer investments cannot obscure the fact that residents vigorously shared in exploiting the mineral resources of their region. Two mining engineers, Samuel F. Emmons and George F. Becker, compiled important statistics on this phenomenon while studying the precious-metal mining industry for the federal census of 1880. Out of a total of 98.80 percent of the deep mines they reported on in Colorado, 58.17 percent were owned within the state, 40.63 percent by people in other states, and 1.20 percent by foreigners, principally Englishmen; translated into numbers, this meant that 146 of their 248 deep mines showed Colorado ownership. Though the relatively small numbers of mines surveyed, the burgeoning leasing system, and the complexity of financial arrangements probably caused some confusion about actual ownership and control, the Emmons and Becker report credited Colorado with a substantial percentage of home ownership less than twenty-five years after the start of the industry. In 1898, Otto Mears, renowned as a road builder in the rugged San Juan region and himself a mining

[22] Len Shoemaker, "Roaring Fork Pioneers," in *Brand Book of the Denver Posse of the Westerners for 1962*, pp. 58–65; Hall, *History*, 4: 494–95; *EMJ* 70 October 20, 1900): 457–58; *Silver Plume* (Colo.) *Silver Standard*, November 10, 1888.

investor, estimated that three-fourths of all mining in the state was done by local capitalists, due, he reasoned, to easterners selling their properties when silver prices declined and the depression of the 1890s caused retrenchment. "Old mines that were abandoned by their non-resident owners have been purchased, reopened and developed by Colorado men with good results," he boasted. Believing that most of Leadville's mines were owned in Denver, Cripple Creek was controlled by Colorado Springs, and the mines of the northern counties were in the hands of Denver financiers, Mears concluded that "nearly all" mining profits stayed within the state. Three years later, the *Mining and Scientific Press* declared that Colorado stood alone in the mining West for paying nine-tenths of its mining dividends to local people.[23]

Cooperating with outside capitalists, learning from their failures and successes, and benefiting from their financial penetration of the Rocky Mountain frontier in the nineteenth century, Colorado had begun to develop its own business leaders capable of owning and operating the mines and guiding the state's economy in the future. Yet, if at any time up to the 1890s there had been a sudden and complete desertion of the mines by absentees, this growth would have been crippled and perhaps brought to a grinding halt, for it was primarily because nonresidents exploited the mineral resources that Coloradans could do the same. Moreover, eastern money committed to developing the high-risk gold and silver mines of the West liberated local people from total dependence on an erratic industry and at the same time enabled them to build enterprises with a more stable and permanent future. Thus, while a Cyrus McCormick or a Norvin Green calculated the odds in a speculative mining venture, David Moffat or John Campion could consider the acquisition of real estate or bank shares, a choice they might never have had without the eastern financier. Rather than smothering local initiative, the ties that bound an undeveloped economic frontier to the capital exporting regions of the East released the energies of local men and helped to mold an autonomous group of entrepreneurs in the mountains of Colorado.

[23] Among the 102 absentee-owned mines, 68 were held in New York, 9 in Illinois, and 5 each in Massachusetts and Missouri (Samuel F. Emmons and George F. Becker, *Statistics and Technology of the Precious Metals*, U.S. Dept. of the Interior, Census Office, pp. 111–13). Mears quoted in *MSP* 76 (May 14, 1898): 518–19. Also, *MSP* 82 (January 19, 1901): 96.

7

In Retrospect

"'I can't help laughing,' " said the son of a wealthy Cleveland merchant who had lost $20,000 on a San Juan silver mine, " 'when I think of what a d——d good thing we would have had if we had only struck it.' " His spirits still high, despite the loss of a large sum of money, this one easterner reflected the sporting mood of many mining investors during the last four decades of the nineteenth century who gave in to the irresistible pull of the western gold and silver mines, only to discover the harsh reality that mining could cost more than it made. Regrettably, the feverish, reckless tempo with which Americans chased the chimera of extracting immense fortunes from the Rocky Mountains makes it difficult to judge the profitability of investments in the precious-metal mines. For one reason, the "unlucky never tell their misfortunes," as one Wall Street speculator observed. What records we do have are frequently marred by rumor, error, innuendo, and omissions, which in their own peculiar way illustrate the condition of the industry but do not provide adequate evidence to measure in full the success of absentee investors. Remember also that the mines were regarded as sheer gambles—always being compared to faro, then the most popular and notorious game of chance in the country—and thus few respectable men of business wished to share with the public their weakness for the flip of the card, the roll of the dice, or, for that matter, the vice of mining speculation. Likewise, a variety of promoters, journalists, and local boosters added more heat than light to the business by their perfervid claims that the mines produced instant millionaires and by their warnings to all that if you can't "boost, don't knock." Obviously, croakers were not tolerated by publicists who hoped to pick eastern pockets for the western mines.

When the *Engineering and Mining Journal* tried to steer investors away from a dubious scheme in 1879, the prime mover of the project angrily retorted that it ought to change its name to "The Sneering, Jeering and Undermining Journal." Consequently, a fuller appraisal of eastern investment in the Colorado mines must look beyond the myths and discount the ballyhoo that infected the industry.[1]

That vast sums of eastern money reached the Colorado mines, even accounting for the sizable amounts of good investment capital siphoned off by fraudulent schemes to trap the uninitiated, is indisputable, and every mining camp in the Rockies had its absentee-controlled companies. Making the short journey between the relatively minor camps of Pitkin and Tin Cup in 1896, one reporter noted the presence of eastern capital on all sides, as he passed within the space of ten miles, mining companies owned by New York, New Jersey, Illinois, St. Louis, and Kansas City investors. During a major boom, such as that which focused on Leadville between 1879 and 1881, many millions of eastern dollars poured into the region and dotted every hillside with mining shafts, and, after the fever subsided, fresh capital continued to arrive for new development work and expansion. The Chrysolite, for instance, survived as an absentee enterprise for nearly twenty years after its formation in 1880. Based on an over-simplified analysis of outside interests in Colorado, one historian broadly indicted nonresident capitalists, finding Colorado of the 1890s the "creation of financial imperialism" and the "victim of absentee ownership and control." Easterners did finance most of the deep mines, did enjoy the profits—if there were any—did branch out into a multitude of new enterprises, and did wield an important influence over the state's social and economic progress. Several of these characteristics can be seen in New York merchant Jerome B. Wheeler. He began by investing in a mine and smelter at Aspen but soon enlarged and diversified his interests to include coal mines, the Colorado Midland Railroad, Wheeler banks in Aspen and Manitou Springs, glass works that made pickle jars and whisky flasks, and a mineral-water

---

[1] Alice Polk Hill, *Tales of the Colorado Pioneers*, p. 201; James K. Medbury, *Men and Mysteries of Wall Street*, p. 196; Henry B. Clifford, *Rocks in the Road to Fortune or the Unsound Side of Mining*, p. 323; *Mining Record* 6 (October 18, 1879): 305.

firm that supplied New York hotels with Colorado seltzer and ginger champagne. The strait-laced Chicagoan John V. Farwell, because he owned a huge placer mine near Hahn's Peak in the mid-1870s, was able to prohibit saloons and gambling in the nearby settlement of Bugtown. Signs of this kind revealed the impact eastern capital had on the Rockies.[2]

However, the facile assumption that eastern investors earned quick and easy profits in the Colorado mines is a myth, and the idea probably stemmed from the unrestrained enthusiasm of promoters, rather than the unvarnished truth about hardrock mining on the frontier. The costs of doing business in the Mountain West were exceptionally high and the hazards great and persistent, facts confirmed by the derelict remains of absentee companies that littered every mining camp in the Rockies. Instead of supplying an endless stream of profits, as promised by the boosters, many mining ventures became bottomless pits for wasted expenditures that wore away at investor patience and bank accounts. Having toured Gilpin and Clear Creek counties in 1870, Californian Almarin B. Paul remarked that Colorado was a very promising mining territory, but "she had been much more of an absorber than a developer" of money and that it requires a "great deal of hope to run most of the mines." A former Cripple Creek miner remembered that in 1900 it was "generally accepted" that five dollars were invested for every dollar made from the mines, and this condition was seen by local men as a definite improvement over 1898, when the ratio was seven dollars for each one extracted from the earth.[3]

In spite of the natural reluctance of industry promoters to dwell on the problems of profitability, western mining men quickly shed these inhibitions when their economic interests seemed immediately

    [2] Leon W. Fuller, "Colorado's Revolt Against Capitalism," *Mississippi Valley Historical Review* 21 (December, 1934): 345; *EMJ* 39 (April 25, 1885): 277; *EMJ* 67 (May 27, 1899): 626; Carlyle C. Davis, *Olden Times in Colorado*, p. 327; Frank Hall, *History of the State of Colorado*, 3: 373; Marshall Sprague, *Newport in the Rockies*, p. 113; E. Shelton, "The Religious Side of Pioneering in Routt County," *Colorado Magazine* 7 (November, 1930): 235; Thomas Tonge, "The Less-Known Gold Fields of Colorado," *Engineering Magazine* 11 (1896): 1037–38.
    [3] *MSP* 20 (April 9, 1870, and April 16, 1870): 234, 250; Leo J. Keena, "Cripple Creek in 1900," *Colorado Magazine* 30 (October, 1953): 270.

threatened. This was the case in 1863 and 1864 when Congress, need-
ing money for the war effort and aware of reports from the West
touting the mines as limitless storehouses of precious metal, debated
fixing a 5 percent tax on the gross yield of the gold and silver properties in the territories. A staunch defender of the infant industry
soon appeared in Sylvester Mowry, an Arizona miner, who argued in
several letters to New York newspapers that mining was "not always
the successful business it is generally represented to be," that it requires enormous sums of capital, and that it often impoverishes stockholders. Though Congress was considering this tax measure in the
midst of a great Colorado stock boom, the *Mining and Scientific Press*
curiously expressed surprise that Washington politicians possessed
false information about the profits of mining and apparently knew
nothing about its actual expenses and losses. Unnecessarily fretful, the
journal moaned that it was "almost an impossible accomplishment to
make our national representatives from other states comprehend, in
a rational manner, the true relation of business and facts, in connection with our mining industry."[4]

Again, in the winter of 1893, shortly before the collapse in the
price of silver, which had been declining since the mid-1880s, leaders
of the industry, fearing economic strangulation, exposed the canard
that deep mining was a glamorously profitable enterprise. Clearly
meaning to promote the free coinage of silver, Senator William M.
Stewart of Nevada included the testimony of several prominent Coloradans in a report prepared by his Committee on Mines and Mining
in February that showed the cost of producing silver was higher than
its market value, then hovering around eighty-seven cents an ounce.
Essentially, the investigation disclosed regular and heavy losses of
capital invested in the industry, and, excluding pure speculation and
wildcat operations, mining, according to Stewart, was "prosecuted at
a greater average loss than is suffered by any other business in the
United States, and that it would not be prosecuted at all if it was not
for the occasional discoveries from which large fortunes are rapidly
acquired." In reaching this conclusion, he had considered the cost to

[4] Mowry wrote his letters to the *New York World* and the *New York
Herald* in April and May, 1864 (collected in Sylvester Mowry, *The Mines of the
West*, pp. 5–6, 9–11). *MSP* 7 (December 21, 1863): 1; *MSP* 8 (May 7, 1864):
308.

investors of locating, opening, developing, and equipping the mines, the many marginally productive ones as well as the few bonanzas.[5]

Horace Tabor, Nathaniel Hill, lawyers Charles S. Thomas and Thomas M. Patterson, David M. Hyman, and journalist Carlyle C. Davis of Colorado willingly participated in the report. With varying degrees of ardor and dramatic effect, each offered a bit of evidence revealing the unhappy state of silver mining due to the plummeting price of the metal, and several took the opportunity to point out the financial insecurity of the industry. Patterson, who claimed he had handled 50 percent of the Colorado mining cases that went before the Supreme Court, frankly admitted that the industry was primarily a high-stakes gamble in which one mine paid to a score that did not, though investors continued to spend in hopes of hitting a jackpot bonanza. The sharpest statement came from W. F. Reinert, editor of the *Colorado Daily State Mining Journal* of Denver, who repeated a common expression in the Rockies that " 'for every dollar taken out of the ground $100 have been put in,' " adding that

this undoubtedly is an over estimation or exaggeration, but it is an indisputable fact that the amount of money actually invested in mining every year exceeds the profits arising from such investment many times. Every mining investor knows, or should know, that the chances are always against him on general principles, and that the chances in his favor always exist, or nearly so, in the bonanza mine, that in most cases only becomes a bonanza after an immense amount of money has been spent on it. This money is always expended before its great value becomes a matter of public intelligence; hence, the public only see the immense profits and are not cognizant of the immense amount of capital that it has taken to bring the property into bonanza.

The director of the United States Mint informed Stewart that the "great majority" of silver mining enterprises "fail to yield any profit on the outlay and generally exhaust the capital employed in their development." Elsewhere, engineer James D. Hague rejected as erroneous the idea that the overall profitability of silver mining could be judged by a small number of highly efficient, bonanza mines. "It is, in fact, only the fortunate finder of a bonanza who gains any notable

[5] U.S., Congress, Senate, Committee on Mines and Mining, *Cost of Production of Gold and Silver Bullion*, 52nd Cong., 2nd Sess., 1893, pp. 2–4.

profit in mining the precious metals; it is almost always only in the hope of finding such prizes that people engage in the business," and to assess the condition of mining on the basis of few successes" is about as near the mark as it would be to take the fortune of the lucky winner of the grand prize in the Louisiana Lottery as a measure of the success of the average lottery speculator." Aware of the rich rewards habitually promised by promoters, one mining journal lamented that such confessions would "scarcely encourage investment in silver mining properties."[6]

Staggered by collapsing silver prices in 1893, spokesmen for the industry strongly stressed the costs and hardships of mining, hoping to get some relief from the United States Treasury in the form of new silver purchases. Even at times when better conditions prevailed there were signs that more millionaires may have gone into the mining business than ever came away from it. In 1878, William Weston, an English-born promoter in the San Juan region, declared that most people, seduced by the dream of instant wealth, had "put more silver into the mine than they ever get out of it," reminding him of the "old Yorkshireman who said that when he and his wife commenced business they ate porridge and after a while chicken, but when his son commenced business he wanted to commence with the chicken and consequently failed." A Denver paper in the 1880s expressed surprise at the successful investments made by Jerome B. Wheeler at Aspen and remarked that eastern capitalists were "apt to falter on the rapid melting away of the coined dollar in its search for the uncoined bullion." The huge fortunes acquired by a "few fabled mining millionaires," said the United States Census Office in 1890, had been offset by the enormous amount of money spent on the mines, putting the average cost of mining gold and silver far above their refined values. Reminiscing about a lifetime in the Colorado mining districts, journalist Carlyle Davis believed he could name every man who made money in mining in two minutes, but he knew thousands who had been losers.[7]

[6] Ibid., pp. 8, 15, 23; James D. Hague, "The Cost of Silver and the Profit of Mining," *The Forum* 15 (March, 1893): 61; *MSP* 66 (February 25, 1893): 114–15.

[7] For another view of the unprofitableness of mining, see Franklin H. Head, "Need of an International Monetary Agreement, *The Forum* 17 (March–August,

Where did the millions of dollars spent by absentee capitalists go, especially if most investors did not reap a rich harvest from the mines? At best only partial answers can be given to a question that requires painstaking research into the records of hundreds of enterprises (most of which left the barest evidence of their existence) and an analysis of individual mining camps, metal-market prices, and the quantity and quality of ore production. Moreover, the mines differed so greatly in regard to location, geology, management, and technology that they beggar many generalizations. T. A. Rickard, describing the Cripple Creek district, provided a helpful clue to understanding the wide diversity of the mining business in the West.

Mines, like men, must be exhausted at last; some men lose their vigor long before old age, many die as infants. It is so with mines. Every mine must have a horizon at which it is at its maximum productiveness. What that horizon is depends upon the particular conditions which obtain in each instance. Some ore bodies are like newborn infants and die as soon as they are chronicled. Many last long enough to enrich one owner and then to delude his successor. Others persist with masterful strength to a great depth and threaten to test the utmost limits of human ingenuity in the pursuit of them underground.

Nor did it help to know that a district contained "dividend-paying" mines, since the earnings paid to shareholders did not necessarily indicate profits, and on that difference often hinged the long-term financial success or failure of a company.[8]

The boom psychology that infected people whenever they considered the mineral resources of the West defined success in mining as recovering the largest possible fortune in the shortest possible time. Mining stimulated the nerve ends of American greed. With careless disregard for the natural resource and the long-term health of their

1894): 458–59. Weston in *EMJ* 25 (March 2, 1878): 151; *Denver Inter-Ocean*, October 25 [1884], quoted by Frank L. Wentworth, *Aspen on the Roaring Fork*, p. 242; U.S. Dept. of the Interior, Census Office, *Report on Mineral Industries in the United States at the Eleventh Census: 1890*, p. 149; Davis, *Olden Times in Colorado*, p. 314.

8 *MSP* 70 (March 9, 1895): 146; Samuel S. Wallihan and T. O. Bigney, *The Rocky Mountain Directory and Colorado Gazetteer for 1871*, p. 167; Thomas A. Rickard in Fred Hills, *The Official Manual of the Cripple Creek District, Colorado, U.S.A.*, 1: 28.

own pocketbooks, mining companies stripped away the richest ore to pay huge and immediate returns to stockholders, while they let the mines slip into appallingly poor shape, foregoing steadier earnings for a splendid moment of wealth. As early as 1867 the *Mining and Scientific Press* found that American companies "generally pick out the eyes of a mine as soon as they get within reach of them, and then usually leave the mine to darkness and abandonment." Consequently, the life span was very brief, often five to seven years, during which time the highest grade ore was excavated at a fast and furious pace, development work slighted, and big profits declared simultaneously with lavish expenditures. The celebrated mining geologist Clarence King observed in 1885 that the "desire for immediate returns has been carried to an extreme and has entailed unnecessary loss. To open mines rapidly usually requires a greater outlay than where work is prosecuted at a slower rate." Asserting that Europeans would have developed the mines more systematically, carefully, and economically, King nonetheless excused the American practice, since the "money necessary to prosecute dead work, depended upon the speculative interest, which looked to brilliant and speedy discoveries of great bonanzas, and that this interest would hardly have supported the delay which would have resulted from a system more economical but less suited to the time and the locality."[9]

Wasteful practices and high costs typified the Colorado industry in the nineteenth century in part because absentee investors succumbed to the erroneous idea that the mines were so rich they could easily afford large expenditures. When mining engineer John Murphy instructed capitalists in the need for caution and prudence in mining, they supposedly looked him in the eye "in a manner indicating the existence of insane hospitals." One major element in the costs to mine

---

[9] Mining men treated the forests no more sparingly. Rossiter W. Raymond found them guilty of "culpable carelessness" in cutting down whole mountainsides of pine and using only the choicest specimens in the mines. In some cases, mining companies were surrounded by damaged and rotting trees while they imported lumber at great expense (*Statistics of Mines and Mining in the States and Territories West of the Rocky Mountains* [1870], p. 343). Clarence King in the Introduction to Samuel F. Emmons and George F. Becker, *Statistics and Technology of the Precious Metals*, U.S. Dept. of the Interior, Census Office, pp. vii–ix; Thomas A. Rickard, ed., *The Economics of Mining*, pp. 158–59; *MSP* 15 (August 3, 1867): 75; *MSP* 55 (December 31, 1887): 421; *MSP* 73 (October 3, 1896): 275; John H. Tice, *Over the Plains, on the Mountains*, p. 228.

owners was milling and smelting the ores. In the 1860s, the primitive stamp mills constructed around Central City and Black Hawk saved only a portion, frequently less than 50 percent, of the precious metal, while the rest was flushed away with the worthless slimes, causing one critic to hear the waters giving a "gleeful laugh as they bear away treasures stolen from the rock." The introduction of the smelting process solved many of the difficulties in reducing refractory ores and increased the rate of return on the metal content, but smelting was itself expensive and trimmed away some of the vaunted profitability of mining. Significantly, the richest ores alone could bear the cost of smelting, and the leaner grades had to be left in the mines or thrown on the dumps to await improved technology and lower rates. Ore removed from the Caribou mine in the early 1870s had to contain eighty ounces of silver to the ton in order to be profitable, and no less than sixty ounces to the ton was profitable at Leadville in 1879. A Chicago visitor to Durango in 1892 believed that several prosperous smelting men in the district could well say, " 'Let me smelt the ores and I care not who owns the bonanzas.' "[10]

Until the railroad arrived in the mining regions and such mechanical devices as the electric-powered tramway were developed, transporting ore to the mills on burros and in horse-drawn wagons was often a "depressing item of expenditure" for mining companies. Jerome B. Chaffee complained about the "terrible expense" of freighting in the early days of Leadville, while Rossiter Raymond pointed out that it cost ten to fifteen dollars a ton to carry ore down the mountains at Georgetown in 1870. These conditions, which had a parallel in every mining camp situated in the Rockies, meant that only the richest ore could be transported at a profit. By the same token, goods and machinery destined for camps tucked away in the far reaches of the mountains usually multiplied in costs as a result of the difficulties encountered in hauling them over the perpendicular terrain in wagons or precariously strapped to the backs of mules. In one case a heavy load of machinery for a mine at Rico took sixty-six days to make the trip from Alamosa (about 120 miles away) at a cost of fifteen cents

---

[10] G. W. Baker, *Treatment of Gold Ores in Gilpin County, Colorado*, pp. 2–3, 9; Tice, *Over the Plains*, p. 226; M. L. Scudder in *The Economist* 8 (August 27, 1892): 297; John G. Murphy, *Practical Mining*, p. 92; Davis, *Olden Times in Colorado*, p. 189; Rickard, ed., *Economics of Mining*, p. 216.

a pound. Given the impact of these expenses on the industry, the Chicago promoter Amasa McCoy informed his clients that the "ore of a mine may be exceedingly rich as to quality and very abundant as to quantity, and yet the expense of production may be such, that every hundred dollars will cost one hundred and one." [11]

Labor, skilled and unskilled, represented another major item in the costs incurred by mining companies. Surely, high wages proved more successful in attracting and keeping workers at remote and hazardous mining operations than did one ludicrously sentimental picture of the miner at work presented by a Georgetown newspaper in 1877.

The field of his labor is under the ground, far away from the sight of man. He had a nice cool place to work, where no proud passer-by can have a chance to scornfully look at him as an inferior being. The work he has to do, when his hands are skillfully trained, is easy, and there is an excitement about it that makes each day interesting. If the vein is large and rich he feels conscious of his importance, for he knows he is unlocking nature's strongest safe and taking therefrom coin for all people and adding to the wealth of the world. If the crevice happens to be narrow and the ore poor he works with the hope of its getting better. Therefore whether in "poverty's vale or abounding in wealth" his labor is lightened by a pleasing excitement which hastens the close of day and cheers him on to the end of the week.

In the forty years prior to 1900, the foreign immigrants and unlucky prospectors who formed the majority of hardrock miners earned on the average $2.50 to $3.50 a day, local conditions and the time period causing some variations. Multiplied by dozens and even hundreds of men employed in some of the larger enterprises, the bill for labor often reached very great proportions. In the mid-1880s, Samuel Emmons and George Becker discovered that 177 deep mines in Colorado

---

[11] According to one source, forty-six mule trains, costing $12,000, were required each day to haul silver bullion from Leadville to the nearest railroad (*Mining Record* 6 [September 6, 1879]: 182). Thomas A. Rickard in Richard P. Rothwell, ed., *The Mineral Industry, Its Statistics and Trade, In the United States and Other Countries to the End of 1894*, p. 644; Jerome Chaffee in *MSP* 37 (November 2, 1878), 284; Raymond, *Statistics of Mines* (1870), p. 376; I. Bonner, "Leadville," *Lippincott's Magazine* 24 (November, 1879): 605; Emmons and Becker, *Statistics and Technology*, p. 170; Carl F. Mathews, "Rico, Colorado—Once a Roaring Camp," *Colorado Magazine* 28 (January, 1951): 41; Amasa McCoy, *Mines and Mining in Colorado: A Conversational Lecture*, p. 12.

supported an aggregate yearly payroll of $3,663,827, or more than $20,000 a company, and this figure did not reflect people employed in managerial positions. When several companies in Gilpin County complained about paying wages of $18 to $21 a week in 1870, the *Central City Herald* agreed that wages were entirely too high and quipped that the mine owners worked for "grandeur," while the miners worked for "money."[12]

Some people who studied the western mining industry concluded that the large expenditures made by companies for supplies, labor, and services contributed more to the local economy than absentee capitalists took away in profits from the mines. The time-honored adage that it takes a mine to run a mine revealed a basic truth: the profits (and losses) in mining ran both ways—not only did investors exploit the mineral resources of an undeveloped region, but Colorado also took advantage of the capital resources of the East. Samuel Emmons and George Becker asserted that the multitude of mining companies that failed to pay their nonresident owners had, nonetheless, been profitable to the West. A mine paying wages and purchasing supplies locally, they reasoned, certainly paid "profits" by supporting the region in much the same way as subsistence farming. Consequently, "the wages of miners are among the profits of the mining industry; so, too, is the surplus over actual cost of the industries which depend upon the mines for a market and that of the industries which depend upon the mines for supplies." According to the *Silver Cliff Prospect* in 1880, four large eastern-owned mining companies in the district employed over five hundred men and spent $90,000 each month for labor and supplies, and the newspaper estimated that 50 to 75 percent of the money made from the mines stayed in the vicinity. Likewise, between 1887 and 1889, the Pay Rock Mining Company of Georgetown reportedly spent $85,000 for labor, $20,000 for supplies, and $40,000 for machinery in the town and neighboring communities. And three banks at Central City in 1880 held accounts for mining laborers totaling nearly three-quarters of a million dollars. Aware of

---

[12] Emmons and Becker, *Statistics and Technology*, p. 157; Frank Fossett, *Colorado, Its Gold and Silver Mines, Farms and Stock Ranges, and Health and Pleasure Resorts*, p. 586; *Georgetown Miner* quoted in *MSP* 35 (July 14, 1877): 23; *MSP* 47 (July 7, 1883): 4; *MSP* 72 (January 11, 1896): 29; *Central City Herald* quoted in *EMJ* 9 (June 14, 1870): 373.

these aspects of outside investment, the *Mining and Scientific Press* invited more eastern companies to the West in 1901 because they left behind nine dollars for every one they took away, leading it to say in the following year that there was "probably a less percentage of this 'absentee landlordism' in mining than in any other occupation."[13]

Yet the wealth that the Colorado mines shared with investors, promoters, laborers, smelters, and the railroads was limited, and eventually mine owners had to reckon with the inescapable fact that their ore was diminishing or becoming too expensive to extract. The deeper they penetrated into the mountainsides, the more costly it became to hoist the ore, timber and ventilate the shafts, and remove the water that gushed into the workings from numerous underground sources. At times the size of the water problem alone was prodigious. A mine near Idaho Springs in the mid-1890s removed forty tons of water for every ton of ore, the Free Silver mine at Aspen in 1897 pumped 4 million gallons every twenty-four hours (enough to supply a city of 46,000 people), and the Leadville district in 1900 handled 15 million gallons each day, equalling about twenty-nine tons of water raised for every ton of ore. To mines in the peak of prosperity, these expenses might have been counterbalanced by huge blocks of high-grade ore, but at greater depths the leaner ore failed to meet spiraling operating costs. By the turn of the century, the mining industry faced conditions that forbade continued high costs and inefficiency.[14]

The 1890s saw the rise of militant unionism among the hardrock miners, particularly in the activities of the Western Federation of Miners, whose strike against the mines and mills of Cripple Creek in 1894 shut down much of the new gold camp and touched off a decade of labor strife. Rejecting an attempt by the mine owners to extend the working day without increasing wages, the union engaged the com-

---

[13] Emmons and Becker, *Statistics and Technology*, p. 120; *Silver Cliff Prospect* quoted in *MSP* 41 (October 9, 1880): 230; *MSP* 59 (December 21, 1889): 469; *MSP* 82 (January 12, 1901): 34; Fossett, *Gold and Silver Mines*, p. 298; Albert Williams, Jr., "Investments in Mining Properties," *Engineering Magazine* 1 (September, 1891): 739–40.

[14] Frank M. White, "The Gold Mines of Colorado," *Harper's Weekly*, June 9, 1894, p. 539; *MSP* 75 (October 30, 1897): 407; U.S., Bureau of the Mint, *Report of the Director of the Mint upon Production of the Precious Metals in the United States for 1900*, p. 118.

panies in a bitter struggle. When violence occurred between the strikers and the owner-dominated sheriff's posse, Populist Governor Davis H. Waite intervened and helped the union to victory under the watchful eye of the state militia. Two years later, however, the union suffered a setback at the hands of Leadville operators, ailing from the depressed market for silver, who stoutly resisted the miners' organization and its demand for higher wages. Though the mine owners won this battle, which lasted from June, 1896, to March, 1897, it was a short-lived and pyrrhic victory. Many of the mines closed down, their lower levels quickly filling up with water, while outside capital shied away from the beleaguered camp, and still the owners had not destroyed unionism. It became more militant and continued to grow, with over forty locals and more than 48,000 members of the Western Federation in the state by 1902. Apparently, the miners were unwilling to heed the advice of the antiunion *Engineering and Mining Journal* when it called upon everyone in 1893 to cooperate in saving the Colorado mining industry by "a little sacrifice here and a little there."[15]

Labor turmoil coincided with the Panic of 1893 and the start of a ruinous nationwide depression that had a twofold impact: the silver mines suffered a precipitous fall in prices that left the white metal selling for sixty-two cents an ounce in 1894, and the entire industry experienced a sharp decline in new capital investment. The New York mining-stock market in 1894 became "far quieter than the grave, a thousand times duller than ditch water, and far more uninteresting than a report of the Department of Agriculture." While they turned to the federal government for relief and appealed to investors not to sell their mining stocks, Colorado mine owners quickly halted production and reduced their work forces. David H. Moffat alone dismissed 2,200 employees—700 men at Leadville, 300 in Creede, 350 in Rico, and 250 in Cripple Creek. More importantly, the onset of the depression forced the industry to examine itself and to acknowledge some long-standing misconceptions about mining. The *Engi-*

[15] Stewart H. Holbrook, *Rocky Mountain Revolution*, pp. 73–83; *EMJ* 62 (September 26, 1896): 301; *EMJ* 64 (November 13, 1897): 586; N. E. Guyot, "Cripple Creek: An Inside Story," *Engineering and Mining Journal-Press* 118 (December 20, 1924): 966; Melvyn Dubofsky, "The Leadville Strike of 1896–1897: An Appraisal," *Mid-America* 48 (April, 1966): 103, 117–18.

neering and Mining Journal, while admitting that the western "mind
does not retain impressions long," declared that future success rested
on developing the lower-grade ores, greater economy in ore removal
and reduction, and in ridding the industry of promotional puffery
that everyone makes money in mining and that the precious metals
are unaffected by market conditions. "Let the era of romance go,"
said the journal, "and the clear headed common sense and straight
business principle will command success." The message was hard to
accept after decades of swollen promises and squandered money.[16]

Furthermore, two vitally important allies of the mining industry,
the railroads and the smelters, which the miners had wisely and tire-
lessly encouraged throughout the early history of Colorado, proved to
be doubtful servants by the turn of the century. Previously, in the
bonanza days when the mines were young, rich, and near the surface,
the fees charged by the smelters and railroads mattered less to mining
men than bringing these essential subsidiary industries into the mountains. But with the economic hardships of the 1890s, the boosters
turned to grumblers, complaining about discriminatory rate structures
and accusing the railroads and smelters alike of bad faith and ill-will.
From the standpoint of the mines, the railroads now represented a
malevolent force in their midst that overcharged them for carrying
ore to the mills and transporting supplies into the camps—a deeply
distressing specter, since the railways were the lifeline of most Rocky
Mountain mining towns. Feeling a sense of helplessness, some mine
owners charged the railroads with rate discrimination or condemned
them for failing to aid the mines when lower ore values and increased
working costs threatened the survival of the industry.[17]

The economic noose around the neck of the mining industry
seemed to tighten in 1899 with the formation of the American
Smelting and Refining Company, a giant corporation that was capi-

---

[16] Frank L. Wentworth said of Aspen in 1893–1894: "A prosperous mining
camp is not an ideal place for people to learn to practice economy for the spirit of
a mining camp encourages free spending and extravagant expectations" (*Aspen,*
p. 314). *New York Times,* June 30, 1893, p. 1; *EMJ* 56 (November 11, 1893,
November 18, 1893, and November 25, 1893): 492, 516, 539; Hall, *History,* 4:
393.

[17] *MSP* 59 (October 26, 1889): 324; *MSP* 73 (July 4, 1896, and August 8,
1896): 11, 107; *MSP* 76 (January 1, 1898): 11; *EMJ* 64 (December 11, 1897):
704; Western Colorado Congress, *Western Colorado and Her Resources,* pp. 58–59.

talized at $65 million and by 1902 controlled nearly all of the ore buying and reduction business in Colorado. Generally, the "smelter trust" aimed at eliminating the wasteful competition among many small, independent operators and raising the market price of the precious metals by cutting back on ore production, but it was also an outgrowth of the larger combination movement then sweeping American industry. Once the properties of the Guggenheim brothers joined the trust in 1901, its domination was virtually complete and the miners began to feel the pressure of new and strong constraints. More than ever before, mine owners bore the risks of the metal market by paying higher rates for reduction, and, much worse, the supposed "freedom" of mining was passing away.[18]

Though mine owners talked of forming a "mine trust" to fight "this matter of hiring our labor from a labor trust and selling our ores to a smelter trust, by which we are between the upper and nether millstone," attempts at combination fared poorly due to the large number, diversity, and distribution of mining companies and perhaps, as one Coloradan said, to the belief that consolidation was "a menace and an injury to our free institutions." Thus, meeting with only scattered resistance from the miners, the smelter trust raised rates, purchased ore selectively, restricted production, and cooperated with some of the larger railroads to the further disadvantage and outrage of mining men. According to the *Mining and Scientific Press* in 1901, the mine owners found themselves at the mercy of monsters they had created.

From being the people who have employed men to work in the mines, who have employed other men and their transportation service to move ore or metals to points of consumption, and who have employed other men and their beneficiation service to reduce the ore and separate the valuable metal to make it available, they will become very largely the servitors of their original servants.

In 1912, mining engineer Arthur J. Hoskin concurred, stating that the American Smelting and Refining Company had become a "domi-

---

[18] On the birth of the smelter trust, see Isaac F. Marcosson, *Metal Magic: Story of the American Smelting and Refining Company*, pp. 35–69. *MSP* 75 (November 27, 1897): 587; *MSP* 82 (June 8, 1901): 258; *EMJ* 67 (March 11, 1899): 288; *EMJ* 71 (February 16, 1901): 202.

neering factor" in the state's mining industry and that the "decrease of mining in Colorado has been contemporaneous with the oppression of this great corporation."[19]

To the many changes affecting the Colorado mining industry by the turn of the century was added stiffer competition for the investment dollar of absentee capitalists. In gold and silver mining alone, several major discoveries captured public attention, such as those in the Klondike, Mexico, South Africa, and Tonopah and Goldfield in Nevada. Capitalists who still retained an interest in mining also began to seek the huge deposits of low-grade ore that were proving more steadily profitable as technology advanced rather than the hidden and erratic seams of spectacularly rich ore that had once mesmerized eastern investors. Moreover, metals that largely had been ignored and even damned by miners as worthless nuisances—such as copper, lead, and zinc—were yearly growing in value and usefulness to a host of new industries in the United States and Europe. For these reasons and others, financiers had begun to turn away from the gold and silver mines of Colorado for the opportunities awaiting them in a diversifying economy and a widely expanded investment market.[20]

The glittering romance of hardrock mining in Colorado was fast disappearing amid the realities of its risks, high costs, and true profits, and in the realization that it was essentially a business like most others which demanded sound management and full-time attention. The waste, extravagance, and chicanery that marked the earlier years had no place in an industry that now rested in low-grade ores and slim margins of profit. Wherever the boom psychology persists that there is no end to mineral wealth, the history of the Colorado mines should be instructive.

[19] "Beneficiation" is the reduction of ores. *MSP* 81 (October 13, 1900): 430; *MSP* 82 (May 18, 1901): 228; *MSP* 84 (April 19, 1902, and April 26, 1902): 214, 236; *EMJ* 71 (May 11, 1901): 584–85; Thomas A. Rickard, *Retrospect: An Autobiography*, pp. 64–65; Arthur J. Hoskin, *The Business of Mining*, p. 212.

[20] *MSP* 81 (November 10, 1900): 516.

Bibliography

MANUSCRIPTS

Adams, Charles Francis, Jr. Papers. Massachusetts Historical Society, Boston.

Adelberg and Raymond. Papers. Manuscript Division, New York Public Library, New York City.

Bennett, John. "Mining and Smelting in Colorado," Colorado State Historical Society, Denver. Copy of reminiscence, 1884, Bancroft Library, University of California, Berkeley.

Copeland, John M. "Diary," Georgetown, 1876. Colorado State Historical Society, Denver.

Dun and Bradstreet. Papers. Baker Library, Harvard University, Cambridge, Mass.

Ellwood, Isaac L. Papers. Western History Research Center, University of Wyoming, Laramie.

Empress Mining Company. Papers. Western Historical Collections, University of Colorado, Boulder.

General Mining Collection. Denver Public Library, Denver, Colo.

Glenn, William W. Papers. Maryland Historical Society, Baltimore.

Green, Norvin E. Papers. The Filson Club, Louisville, Ky.

Hagerman, James J. Papers. Western History Research Center, University of Wyoming, Laramie. Largely typed copies made and donated by John J. Lipsey. Includes the "Memoirs" of James J. Hagerman, June, 1908.

Holt, George H. Papers. Western Historical Collections, University of Colorado, Boulder.

Hyde, Henry B. Papers. Baker Library, Harvard University, Cambridge, Mass.

Hyman, David M. "The Romance of a Mining Venture" (1916). In possession of the family of Donald M. Hyman, New York City.

Leiter, Levi. Papers. Chicago Historical Society. Chicago, Ill.

McCormick, Cyrus H. Papers. State Historical Society of Wisconsin, Madison.

Moffat, David H., Jr. Papers. First National Bank of Denver Archives, Denver, Colo.

Olcott, Ebenezer E. Papers. New York Historical Society, New York City.

Prince, LeBaron Bradford. Papers. New Mexico State Records and Archives Center, Santa Fe.

Pullman, George M. Papers. Chicago Historical Society, Chicago, Ill.

Randall, Jesse S. Papers. Western Historical Collections, University of Colorado, Boulder.

Smith, Eben. Papers. Denver Public Library, Denver, Colo.

Tabor, Horace A. W. Papers. Colorado State Historical Society, Denver.

THESES AND DISSERTATIONS

Barnett, Paul S. "Colorado Domestic Business Corporations, 1859–1900." Ph.D. dissertation, University of Illinois, 1966.

Hough, Charles Merrill. "Leadville, Colorado, 1878 to 1898: A Study in Unionism." Master's thesis, University of Colorado, 1958.

Sheets, Marjorie. "The Promoter." Master's thesis, University of Chicago, 1917.

NEWSPAPERS AND JOURNALS

American Journal of Mining, Milling, Ore Boring, Geology, Mineralogy, Metallurgy, etc. (New York), 1866–July, 1869.

American Mining Gazette, and Geological Magazine (New York), August, 1865.

American Mining Index (New York), scattered issues 1866.

Boston Evening Transcript, January 18, 1913.

Bullion (New York), 1879–1883.

Central City (Colo.) *Daily Miners' Register,* March 19, 1867.

Dawson, Thomas F. "Scrapbooks." Colorado State Historical Society, Denver. Newspaper clippings.

Denver Republican, January 21, 1881.

Denver Post, December 31, 1903.

Denver Times, scattered issues 1882; February 2, 1901; November 6, 1902.

Denver Tribune, scattered issues 1879–1882.

Denver Weekly Times, April 7, 1880.

The Economist (Chicago), August–October, 1892.

Engineering and Mining Journal (New York), 1869–1902.

The Federal Reporter, 1887.

Financial and Mining Record (New York), July, 1888–October 8, 1892.

Mining and Metallurgy (New York), April, 1920.

Mining and Scientific Press (San Francisco), 1860–1902.

"Mining" File Drawer. Colorado State Historical Society, Denver. Contains typewritten copies of articles from various newspapers and journals.

Mining Record (New York), 1879–1880.

Monterey (Calif.) *Daily Cypress*, January 18, 1913.

New York Times, 1859–1902.

Oakland (Calif.) *Times*, December 26, 1901.

Ouray (Colo.) *Times*, scattered issues 1882.

Rocky Mountain News (Denver), September 16, 1874; 1879–1881; scattered issues 1884–1885; April 23, 1909; October 23, 1911.

San Francisco Call, December 25, 1901.

Silver Plume (Colo.) *Silver Standard*, 1887–1888.

GOVERNMENT PUBLICATIONS

Cross, Whitman, and R. A. F. Penrose, Jr. *Geology and Mining Industries of the Cripple Creek District.* U.S. Geological Survey 16th Annual Report. Washington, D.C.: Government Printing Office, 1895.

Emmons, Samuel F. *Geology and Mining Industry of Leadville, Colorado.* Vol. 12 of U.S. Geological Survey Monographs. Washington, D.C.: Government Printing Office, 1886.

————, and George F. Becker. *Statistics and Technology of the Precious Metals.* U.S. Department of the Interior, Census Office. Washington, D.C.: Government Printing Office, 1885.

Fay, Albert H. *A Glossary of the Mining and Mineral Industry.* U.S. Department of the Interior, Bureau of Mines, Bulletin 95. Washington, D.C.: Government Printing Office, 1948.

Hague, James D. "Mining Industries." *Reports of the U.S. Commissioners to the Paris Exposition, 1878.* Washington, D.C.: Government Printing Office, 1880.

Henderson, Charles W. *Mining in Colorado: A History of Discovery, Development and Production.* U.S. Geological Survey Professional Paper 138. Washington, D.C.: Government Printing Office, 1926.

Raymond, Rossiter W. *Mineral Resources of the States and Territories West of the Rocky Mountains.* Washington, D.C.: Government Printing Office, 1869.

————. *Statistics of Mines and Mining in the States and Territories West of the Rocky Mountains.* Washington, D.C.: Government Printing Office, 1870–1877. Issued annually.

Taylor, James W. *Report on the Mineral Resources of the United States East of the Rocky Mountains.* Washington, D.C.: Government Printing Office, 1867–1868.

U.S. Bureau of the Mint. *Report of the Director of the Mint upon Production of the Precious Metals in the United States during the Calendar Year 1887.* Washington, D.C.: Government Printing Office, 1888.

———. *Report upon Production of the Precious Metals in the United States for 1900.* Washington, D.C.: Government Printing Office, 1901.

U.S. Congress. *Biographical Directory of the American Congress, 1774–1927.* Washington, D.C.: Government Printing Office, 1928.

U.S. Dept. of the Interior, Census Office. *Report on Mineral Industries in the United States at the Eleventh Census: 1890.* Washington, D.C.: Government Printing Office, 1892.

U.S. Senate, Committee on Mines and Mining. *Cost of Production of Gold and Silver Bullion.* Report No. 1310, 52nd Cong., 2nd Sess., 1893.

BOOKS, PAMPHLETS, AND ARTICLES

Allied Mines of Imogene Basin, Ouray County, Colorado. New York: Kilbourne Tomkins, 1880.

American Bureau of Mines. *Prospectus.* New York: William C. Bryant & Co., 1866.

Among the Silver Seams of Colorado. Georgetown, Colo.: Georgetown Courier, 1886.

Anker, Moses. *A Hurry-graph of Colorado and Its Silver Mines.* Including prospectus for the "Anker Tunnels and Silver Mines." Baltimore: Jones Bros., 1870.

Annual Mining Review and Stock Ledger. San Francisco: Verdenal, Harrison, Murphy & Co., 1876.

Arkell, MacMillan, and Stewart, comps. *Aspen, Pitkin County, Her Mines and Mineral Resources.* Aspen, Colo.: Aspen Daily Leader, July, 1892.

Aschmann, Homer. "The Natural History of a Mine." *Economic Geography* 46 (April, 1970): 172–89.

Aspen: Her Mines and Mineral Resources. Aspen, Colo.: Aspen Times [1891].

Baker, G. W. *Treatment of Gold Ores in Gilpin County, Colorado.* Central City: Colorado Herald Book and Job Press, 1870.

Balch, William R., comp. *The Mines, Miners and Mining Interests of the*

United States in 1882. Philadelphia: Mining Industrial Publ. Bureau, 1882.

Bartlett, Robert F. "Aspen: The Mining Community, 1879–1893." In *Brand Book of the Denver Posse of the Westerners for 1950*, edited by Harold H. Dunham. Denver: University of Denver Press, 1951.

Bauman, D. *King Carbonate, Leadville, Colorado*. Chicago: Rand, McNally, 1882.

Bishop, George W., Jr. *Charles H. Dow and the Dow Theory*. New York: Appleton-Century-Crofts, Inc., 1960.

Bliss, Edward. *A Brief History of the New Gold Regions of Colorado Territory*. New York: John W. Amerman, 1864.

Bobtail Gold Mining Company of Colorado. New York: n.p., 1864.

Bonner, I. "Leadville." *Lippincott's Magazine* 24 (November, 1879): 604–15.

Borenstein, Israel. *Capital and Output in Mining Industries, 1870–1948*. Occasional Paper, no. 45. New York: National Bureau of Economic Research, 1954.

Bowles, Samuel. *Our New West: Records of Travel Between the Mississippi River and the Pacific Ocean*. Hartford: Hartford Publ. Co., 1869.

Brayer, Herbert O. "Boom Town Banker—Central City, Colorado, 1880." *Bulletin of the Business Historical Society* 19 (June, 1945): 67–95.

Buell, J. S. *Suggestions Relating to the Organization of Gold and Silver Mining Corporations*. Buffalo: Baker, Jones & Co., 1879.

Buford, General N. B. *Reports, etc. on the Property of the Federal Union Mining Co.* Boston: George C. Rand & Avery, 1865.

Buley, R. Carlyle. *The Equitable Life Assurance Society of the United States, 1859–1959*. New York: Appleton-Century-Crofts, Inc., 1959.

Bullion Consolidated Mining Company. Boston: A. Mudge & Sons, 1866.

Bullion Mining Company. *Prospectus*. New York: John W. Amerman, 1864.

Burt, S. W., and E. L. Berthoud. *The Rocky Mountain Gold Regions*. 1861. Reprint ed. Denver: Old West Publ. Co., 1962.

Canfield, John G. *Mines and Mining Men of Colorado*. Denver: The Author, 1893.

Catalpa Mining Company of Leadville, Colorado. *Prospectus and By-Laws*. n.p.: n.p., 1880.

Caughey, John W. *California*. Englewood Cliffs, N.J.: Prentice-Hall, Inc., 1953.

Central Gold Mining Company. *Prospectus*. New York: Styles, Cash & Co., 1866.

Chemical Gold and Silver Ore Reducing Company. New York: n.p., 1865.

Chrysolite Silver Mining Company. New York: David H. Gildersleeve, 1880.

Clifford, Henry B. *Rocks in the Road to Fortune or the Unsound Side of Mining.* New York: Gotham Press, 1908.

————. *Years of Dishonor or the Cause of the Depression in Mining Stocks, also the Remedy.* New York: United Mining and Investment Co. of North and South America, 1883.

Colorado Diamond Tunnel Silver Mining Company of Baltimore City. *Prospectus.* Annapolis: S. S. Mills and L. F. Colton, 1872.

Colorado: Its Mineral and Agricultural Resources. New York: n.p., 1864.

"Colorado: Resources and Attractions of the State." *Bankers' Magazine* 57 (July, 1898): 177–208.

Colvin Gold Mining Company, *Prospectus.* New York: George F. Nesbitt & Co., 1864.

Corbett, Thomas B. *The Colorado Directory of Mines.* Denver: Rocky Mountain News Printing Co., 1879.

Cornell, S. Douglas. *Report on the Condition and Prospects of Gold Mining in Colorado.* Buffalo: J. S. Leavitt, 1863.

Corregan, Robert A., and David F. Lingane, eds. *Colorado Mining Directory.* Denver: Colorado Mining Directory Co., 1883.

Cowley, Charles. *Reminiscences of James C. Ayer and the Town of Ayer.* 3rd ed. Lowell, Mass.: Penhallow Printing Co., [1879].

Crane, Walter. *Gold and Silver.* New York: John Wiley & Sons, 1908.

Creede Camp: The Greatest Mineral Discovery Since the Days of Leadville. 3rd rev. ed. n.p.: n.p., n.d.

Cunningham, Frank. "David H. Moffat . . . Empire Builder." *Tradition* (Detroit) 4 (1961): 5–13.

Cushman, Samuel, and J. P. Waterman. *The Gold Mines of Gilpin County.* Central City, Colo.: Register Steam Printing Co., 1876.

Davis, Carlyle C. *Olden Times in Colorado.* Los Angeles: Phillips Publ. Co., 1916.

Davis, Floyd. *The Mine Investors' Guide.* Des Moines: Western Correspondence School of Mining Engineering, 1909.

Day and Bushnell Mining Company. *The Colorado Gold Mines.* New York: n.p., 1864.

Defiance Silver Mining Company. *Prospectus and By-Laws, Map, Scientific Reports, Letters, etc.* New York: Fisk, Russell & Ames, 1869.

Defoe, Daniel. "Of Projectors." In *The Earlier Life and Chief Earlier Works of Daniel Defoe,* edited by Henry Morley. 1889. Reprint ed. New York: Burt Franklin, 1970.

Dill, R. G. *History of the Arkansas Valley, Colorado*. Chicago: O. L. Baskin & Co., 1881.

Dobson, Charles M. "Mine Salting." *Cosmopolitan* 24 (April, 1898): 575–83.

Dresher, George B. *A Description of Colorado, Leadville, and the Sovereign Consolidated Silver Mines*. Philadelphia: George W. Arms & Co., 1881.

Dubofsky, Melvyn. "The Leadville Strike of 1896–1897: An Appraisal." *Mid-America* 48 (April, 1966): 99–118.

Ellis, Amanda M. *The Strange, Uncertain Years: An Informal Account of Life in Six Colorado Communities*. Hamden, Conn.: The Shoe String Press, Inc., 1959.

Ellis, Elmer. *Henry Moore Teller: Defender of the West*. Caldwell, Idaho: Caxton Printers, 1941.

Farmer, Elihu J. *The Resources of the Rocky Mountains*. Cleveland: Leader Printing Co., 1883.

Fēls, Rendigs. *American Business Cycles, 1865–1897*. Chapel Hill: University of North Carolina Press, 1959.

Finlay, James R. *The Cost of Mining*. New York: McGraw-Hill Book Co., 1909.

Fite, Emerson D. *Social and Industrial Conditions in the North During the Civil War*. New York: Peter Smith, 1930.

Fossett, Frank. *Colorado: A Historical, Descriptive and Statistical Work on the Rocky Mountain Gold and Silver Mining Regions*. Denver: Daily Tribune Steam Printing House, 1876.

——. *Colorado, Its Gold and Silver Mines, Farms and Stock Ranges, and Health and Pleasure Resorts*. New York: C. G. Crawford, 1880.

Foster, Ernest LeNeve. "The Colorado Central Lode, A Paradox of the Mining Law." *Proceedings of the Colorado Scientific Society* 7 (1901–1902): 41–54.

Fowler, William W. *Twenty Years of Inside Life in Wall Street*. New York: Orange Judd Co., 1880. Reprint ed. New York: Greenwood Press, 1968.

Freeland, Francis T. "Mining Leases." *Transactions of the American Institute of Mining Engineers* 25 (1895): 106–12.

Fritz, Percy S. *Colorado: The Centennial State*. New York: Prentice-Hall, 1941.

Fuller, Leon W. "Colorado's Revolt Against Capitalism." *Mississippi Valley Historical Review* 21 (December, 1934): 343–60.

Gandy, Lewis C. *The Tabors: A Footnote of Western History*. New York: Press of the Pioneers, 1934.

Goad, Thomas W. "Gold and Silver Mining in the Rocky Mountains of Colorado." *Journal of the Society of the Arts* (London) 37 (February 1, 1889): 173–80.

Goodspeed, Thomas W. "Joseph Reynolds." *The University Record* (University of Chicago) 7 (January, 1921): 26–42.

Goodwin, Charles C. *As I Remember Them*. Salt Lake City: Salt Lake Commercial Club, 1913.

Graff, John F. *"Graybeard's" Colorado: or, Notes on the Centennial State*. Philadelphia: J. B. Lippincott & Co., 1882.

Greever, William S. *The Bonanza West: The Story of the Western Mining Rushes, 1848–1900*. Norman: University of Oklahoma Press, 1963.

Gregory Gold Mining Company. *Prospectus and Geological Survey and Report*. New York: n.p., 1863.

Gressley, Gene M. "Colonialism: A Western Complaint." *Pacific Northwest Quarterly* 54 (January, 1963): 1–8.

Griswold, Don L. and Jean H. *The Carbonate Camp Called Leadville*. Denver: University of Denver Press, 1951.

Guice, John D. W. *The Rocky Mountain Bench: The Territorial Supreme Courts of Colorado, Montana, and Wyoming, 1861–1890*. New Haven: Yale University Press, 1972.

Guiterman, Franklin. "On the Use, Non-Use, and Waste of the Mineral Resources of Colorado." *Proceedings of the Colorado Scientific Society* 9 (1911): 431–50.

Guyot, N. E. "Cripple Creek: An Inside Story." *Engineering and Mining Journal-Press* 118 (December 13, 1924, and December 20, 1924): 933–37, 965–70.

Hague, James D. "The Cost of Silver and the Profit of Mining." *The Forum* 15 (March, 1893): 60–67.

———. "Mining Engineering and Mining Law." *Engineering and Mining Journal* 78 (October 20, 1904): 629–30.

Hall, Frank. *History of the State of Colorado*. 4 vols. Chicago: Blakely Printing Co., 1889-1895.

Hammond, John Hays. "Suggestions Regarding Mining Investments." *Engineering and Mining Journal* 89 (January 1, 1910): 8–11.

Hayes, Augustus A. "Grub Stakes and Millions." *Harper's Monthly Magazine* 60 (February, 1880): 380–97.

Head, Franklin H. "Need of an International Monetary Agreement." *The Forum* 17 (March–August, 1894): 455–66.

Hill, Alice Polk. *Tales of the Colorado Pioneers*. Denver: Pierson & Gardner, 1884.

Hill, John, Jr. *Gold Bricks of Speculation*. Chicago: Lincoln Book Concern, 1904.

Hills, Fred. *The Official Manual of the Cripple Creek District, Colorado, U.S.A.* Vol. 1. Colorado Springs: The Author, 1900.

Holbrook, Stewart H. *Rocky Mountain Revolution*. New York: Henry Holt & Co., 1956.

Hollister, Ovando J. *The Mines of Colorado*. Springfield, Mass.: Samuel Bowles & Co., 1867.

Hoover, Herbert C. *Principles of Mining*. New York: Hill Publ. Co., 1909.

Horner, John W. *Silver Town*. Caldwell, Idaho: Caxton Printers, 1950.

Hoskin, Arthur J. *The Business of Mining*. Philadelphia: J. B. Lippincott Co., 1912.

Hotchkiss, George W. *Industrial Chicago*. Vol. 5: *The Lumber Interests*. Chicago: Goodspeed Publ. Co., 1894.

Howard, William W. "The City of Aspen, Colorado." *Harper's Weekly*, January 19, 1889, pp. 57–60.

Howbert, Irving. *Memories of a Lifetime in the Pike's Peak Region*. New York: G. P. Putnam's Sons, 1925.

Hoyt, A. W. "Over the Plains to Colorado." *Harper's Monthly Magazine* 35 (June, 1867): 1–21.

Hume, John F. *The Art of Investing*. New York: D. Appleton & Co., 1888.

Hutchinson, William T. *Cyrus Hall McCormick*. 2 vols. New York: D. Appleton-Century Co., 1935.

Ingham, G. Thomas. *Digging Gold Among the Rockies*. Philadelphia: Cottage Library Publ. House, 1881.

J. B. Grant Silver Mining Company. *Prospectus*. Monmouth, Ill.: J. S. Clark & Son, 1881.

Jackson, W. Turrentine. *The Enterprising Scot: Investors in the American West After 1873*. Edinburgh: Edinburgh University Press, 1968.

Jensen, Vernon H. *Heritage of Conflict: Labor Relations in the Nonferrous Metal Industry up to 1930*. Ithaca: Cornell University Press, 1950.

Johnson, Allen, and Dumas Malone, eds. *Dictionary of American Biography*. 20 vols. New York: Charles Scribner's Sons, 1928–1936.

Keeler, Bronson, *Leadville and Its Silver Mines*. Chicago: E. L. Ayer, 1879.

Keena, Leo J. "Cripple Creek in 1900." *Colorado Magazine* 30 (October, 1953): 269–75.

Kemp, Donald C. *Colorado's Little Kingdom*. Denver: Sage Books, Inc., 1949.

Kent, L. A. *Leadville. The City, Mines and Bullion Product. Personal Histories of Prominent Citizens. Facts and Figures Never Before Given to the Public*. Denver: Daily Times Steam Printing House, 1880.

King, Joseph L. *History of the San Francisco Stock and Exchange Board*. San Francisco: The Author, 1910.

King Solomon Mining Syndicate. *Opinions*. Denver: n.p., n.d.

Kip and Buell Gold Company. *Prospectus*. New York: Bogert, Bourne & Auten, 1864.

Kirkland, Edward C. *Charles Francis Adams, Jr., 1835–1915: The Patrician at Bay*. Cambridge: Harvard University Press, 1965.

LaCrosse Mining, Milling, and Power Company. *Prospectus*. Boulder: News and Courier, 1880.

Lavender, David. *The Story of Cyprus Mines Corporation*. San Marino, Calif.: The Huntington Library, 1962.

Lawrence, Benjamin B. "Notes on the Lease or Tribute System of Mining, as Practiced in Colorado." *Transactions of the American Institute of Mining Engineers* 21 (1892–1893): 911–19.

Lee, Mabel Barbee. *Cripple Creek Days*. Garden City, N.Y.: Doubleday & Co., 1958.

Lewis, Oscar. *Silver Kings: The Lives and Times of Mackay, Fair, Flood, and O'Brien, Lords of the Nevada Comstock Lode*. New York: Alfred A. Knopf, 1947.

Leyendecker, Liston E. "Colorado and the Paris Universal Exposition, 1867." *Colorado Magazine* 36 (Winter, 1969): 1–15.

Lingenfelter, Richard E. *The Hardrock Miners: A History of the Mining Labor Movement in the American West, 1863–1893*. Berkeley: University of California Press, 1974.

Lipsey, John J. *The Lives of James John Hagerman*. Edited by Edith Powell. Denver: Golden Bell Press, 1968.

Lowe, T. H. *Colorado: Its Mineral and Agricultural Resources*. Louisville: German & Bros., 1870.

McClure, A. K. *Three Thousand Miles Through the Rocky Mountains*. Philadelphia: J. B. Lippincott & Co., 1869.

McCoy, Amasa. *Mines and Mining in Colorado: A Conversational Lecture*. Chicago: International Mining and Exchange Co., 1871.

McMechen, Edgar C. *The Moffat Tunnel of Colorado: An Epic of Empire*. 2 vols. Denver: Wahlgreen Publ. Co., 1927.

Manning, Jay F. *Leadville, Lake County and the Gold Belt*. Denver: Manning, O'Keefe & DeLashmatt, 1895.

Marcosson, Isaac F. *Metal Magic: Story of the American Smelting and Refining Company.* New York: Farrar, Straus, 1949.

Martin, Joseph G. *A Century of Finance: History of the Boston Stock and Money Markets, January, 1798, to January, 1898.* Boston: The Author, 1898.

Mathews, Carl F. "Rico, Colorado—Once a Roaring Camp." *Colorado Magazine* 28 (January, 1951): 37–49.

Maynard, George W. "Early Colorado Days." *Mining and Scientific Press* 98 (January–June, 1909): 789–92.

Medbury, James K. *Men and Mysteries of Wall Street.* Boston: Fields, Osgood & Co., 1870.

Merchants and Mechanics Silver Mining Company of Baltimore and Colorado. Baltimore: John W. Woods, 1870.

The Mining Trust Company and Its Exchange Building. New York: n.p., 1880.

Mitchell, Wesley C. *A History of the Greenbacks.* Chicago: University of Chicago Press, 1903.

Mowry, Sylvester. *The Mines of the West.* New York: G. E. Currie, 1864.

Mumey, Nolie. *Creede: History of a Colorado Silver Mining Town.* Denver: Artcraft Press, 1949.

Murphy, John G. *Practical Mining.* New York: D. Van Nostrand Co., 1890.

Muzzey, David S. *James G. Blaine: A Political Idol of Other Days.* New York: Dodd, Mead & Co., 1934.

"My Mining Investments." *Lippincott's Magazine* 27 (January, 1881): 86–96.

National Cyclopedia of American Biography. 49 vols. New York: James T. White & Co., 1898–1966.

Nash, Gerald D. "Government and Business: A Case Study of State Regulation of Corporate Securities, 1850–1933." *Business History Review* 38 (Summer, 1964): 144–62.

Newton, Harry J. *Pitfalls of Mining Finance.* Denver: Daily Mining Record, 1904.

———. *Yellow Gold of Cripple Creek.* Denver: Nelson Publ. Co., 1928.

New York *Bullion*, comp. *Bullion: Its Production and Use.* New York: United States Mining Investment Co., 1880.

New York Mining Directory. New York: Hollister & Goddard, 1880.

Nicholas, Francis C. *Mining Investments and How to Judge Them.* New York: Moody Corp., 1907.

———. "The Wrongs and Opportunities in Mining Investments." *An-*

nals of the American Academy of Political and Social Science 35 (1910): 207–16.

Niehaus, Fred R. *Development of Banking in Colorado.* Denver: Mountain States Publ. Co., 1942.

O'Connor, Harvey. *The Guggenheims: Making of an American Dynasty.* New York: Covici-Friede, 1937.

Ohio Consolidated Mining Company. *Prospectus.* New York: David H. Gildersleeve, 1880.

Otero, Miguel A. *My Life on the Frontier, 1864–1882.* New York: Press of the Pioneers, 1935.

Pacific Coast Annual Mining Review and Stock Ledger. San Francisco: Francis & Valentine, 1878.

Paul, Almarin B. *Mine Development: The Basis of Prosperity.* San Francisco: Mining and Scientific Press, 1896.

Paul, Rodman W. "Colorado as a Pioneer of Science in the Mining West." *Mississippi Valley Historical Review* 47 (June, 1960): 34–50.

————. *Mining Frontiers of the Far West, 1848–1880.* New York: Rinehart and Winston, 1963.

Pelican and Dives Mining Company. New York: n.p. [1880].

Peterson, William J. "Joseph Reynolds." *The Palimpsest* (Iowa City) 51 (April, 1970): 169–78.

Pfanner, Robert. "Highlights in the History of Fort Logan." *Colorado Magazine* 19 (May, 1942): 81–91.

Pickering, J. C. *Engineering Analysis of a Mining Share.* New York: McGraw-Hill Book Co., 1917.

Pontiac Gold Company of Colorado. *Reports, etc.* New York: Latimer Bros. & Seymour, 1864.

Popham, Donald F. "The Early Activities of the Guggenheims in Colorado." *Colorado Magazine* 27 (October, 1950): 263–69.

Portrait and Biographical Record of Denver and Vicinity. Chicago: Chapman Publ. Co., 1898.

Portrait and Biographical Record of the State of Colorado. Chicago: Chapman Publ. Co., 1899.

Pourtales, James. *Lessons Learned From Experience.* Translated by Margaret W. Jackson. Denver: W. H. Kistler, 1955.

Prudential Mining Investment Association. *Prospectus.* Florence, Colo.: Daily Herald [1896].

Raymond, Rossiter W. "The Law of the Apex." *Transactions of the American Institute of Mining Engineers* 12 (1883–1884): 387–444.

Rickard, Thomas A. "The Development of Colorado's Mining Industry."

Transactions of the American Institute of Mining Engineers 26 (1896): 834–48.

——. ed. *The Economics of Mining*. New York: Hill Publ. Co., 1907.

——. "Gold Mining Activity in Colorado." *North American Review* 162 (1896): 473–80.

——. "Gold Stamp Milling." *Cassier's Magazine* 5 (November, 1893): 27–33.

——. *A History of American Mining*. New York: McGraw-Hill Book Co., 1932.

——, ed. *Interviews with Mining Engineers*. San Francisco: Mining and Scientific Press, 1922.

——. *Retrospect: An Autobiography*. New York: Whittlesey House, 1937.

——. *The Romance of Mining*. Toronto: The Macmillan Co. of Canada Ltd., 1945.

——, ed. *Rossiter Worthington Raymond: A Memorial*. New York: American Institute of Mining and Metallurgical Engineers, 1920.

——. *The Sampling and Estimation of Ore in a Mine*. New York: Engineering and Mining Journal, 1904.

Ritter, Etienne. *From Prospect to Mine*. Denver: Mining Science Publ. Co., 1910.

Rolker, Charles M. "Notes on a Fire-Bulkhead." *Transactions of the American Institute of Mining Engineers* 13 (February, 1884–June, 1885): 505–10.

Rollins, John Q. A., Jr. "John Q. A. Rollins, Colorado Builder." *Colorado Magazine* 16 (May, 1939): 110–18.

Rothman, David J. *Politics and Power: The United States Senate, 1869–1901*. Cambridge, Mass.: Harvard University Press, 1966.

Rothwell, Richard P., ed. *The Mineral Industry, Its Statistics and Trade, In the United States and Other Countries to the End of 1894*. New York: Scientific Publ. Co., 1895.

Sanford, Albert B. "John L. Routt, First State Governor of Colorado." *Colorado Magazine* 3 (August, 1926): 81–86.

Schuch, Philip, Jr. *A Cripple Creek Gold Mine and the Law*. Cripple Creek, Colo.: Morning Times, 1898.

Shelton, E. "The Religious Side of Pioneering in Routt County." *Colorado Magazine* 7 (November, 1930): 235–41.

Shelton, M. B. *Rocky Mountain Adventures*. Boston: Christopher Publ. House, 1920.

Shoemaker, Len. *Pioneers of the Roaring Fork*. Denver: Sage Books, 1965.

——. "Roaring Fork Pioneers." In *Brand Book of the Denver Posse of the Westerners for 1962*, edited by John J. Lipsey. Denver: Johnson Publ. Co., 1963.

Shuck, Oscar T. *Sketches of Leading and Representative Men of San Francisco.* London and New York: London and New York Publ. Co., 1875.

Silver Cliff Mining Company. *Annual Report, 1880.* New York: n.p., 1880.

———. *Prospectus.* New York: n.p. [1879].

Simonin, Louis L. *The Rocky Mountain West in 1867.* Translated and edited by Wilson O. Clough. Lincoln: University of Nebraska Press, 1966.

Smiley, Jerome C. *History of Denver.* Denver: Times-Sun Publ. Co., 1901.

Smith and Parmelee Gold Company. New York: Sackett & Cobb, 1864.

Smith, Duane A. *Horace Tabor: His Life and the Legend.* Boulder: Colorado Associated University Press, 1973.

———. *Rocky Mountain Mining Camps: The Urban Frontier.* Bloomington: Indiana University Press, 1967.

———. *Silver Saga: The Story of Caribou, Colorado.* Boulder, Colo.: Pruett Publ. Co., 1974.

Smith, Frederick H. *Rocks, Minerals and Stocks.* Chicago: The Railway Review, 1882.

"Speculator, A" *A Speculation* (novel). Denver: D. M. Richards Publ. Co., 1884.

Spence, Clark C. *British Investments and the American Mining Frontier, 1860–1901.* Ithaca: Cornell University Press, 1958.

———. *Mining Engineers & The American West: The Lace-Boot Brigade, 1849–1933.* New Haven: Yale University Press, 1970.

Sprague, Marshall. *Money Mountain.* Boston: Little, Brown & Co., 1953.

———. *Newport in the Rockies.* Denver: Sage Books, 1961.

Spring, Agnes Wright. *The First National Bank of Denver: The Formative Years, 1860–1865.* Denver: Bradford-Robinson Printing Co., 1960.

Standard Gold Company of Colorado. *Prospectus and Report.* New York: n.p., 1864.

Stanton, Irving W. *Sixty Years in Colorado: Reminiscences and Reflections of a Pioneer of 1860.* Denver: Privately printed, 1922.

Stone, Ross C. *Gold and Silver Mines of America.* New York: Scientific Publ. Co., 1878.

Strahorn, R. E. *Gunnison and San Juan.* Omaha: New West Publ. Co., 1881.

Stretch, R. H. *Prospecting, Locating and Valuing Mines.* New York: McGraw-Hill Book Co., 1909.

Stuart, John M. *Mining: Its Theory and Practice.* New York: n.p., 1879.

Taylor, Bayard. *Colorado: A Summer Trip.* New York: G. P. Putnam & Son, 1867.

Taylor, Robert G. *Cripple Creek.* Indiana University Publications, Geographic Monograph Series, vol. 1. Bloomington: Indiana University Press, 1966.

Teetor, Henry D. "Bankers and Capitalists of Colorado." *Magazine of Western History* 11 (November, 1889–April, 1890): 56–60.

———. "The First Quartz-Mill in Colorado." *Magazine of Western History* 13 (November, 1890–April, 1891): 563.

Thayer, William M. *Marvels of the New West: A Vivid Portrayal of the Unparalleled Marvels in the Vast Wonderland West of the Missouri River.* Norwich, Conn.: Henry Bill Publ. Co., 1893.

Tice, John H. *Over the Plains, On the Mountains.* St. Louis: Industrial Age Printing Co., 1872.

Tonge, Thomas. "The Evolution of Mining and Ore Treatment in Colorado." *Engineering Magazine* 18 (1900): 265–76.

———. "The Less-Known Gold Fields of Colorado." *Engineering Magazine* 11 (1896): 1029–43.

———. "The Mining Development of Gilpin County, Colorado." *Engineering Magazine* 22 (1902): 203–19.

United Mines Company. *Prospectus.* n.p.: n.p., n.d.

United States Annual Mining Review and Stock Ledger. New York: Mining Review Publ. Co., 1879.

Upshur, George L. *As I Recall Them: Memories of Crowded Years.* New York: Wilson-Erickson, Inc., 1936.

Vandemoer, H. R., comp. *Diaries, Newspaper Articles, and Letters of John J. Vandemoer.* Denver: The Compiler, 1956.

Van Diest, Peter H. "On the Estimation of the Capital Requisite for Investment in Mining Properties." *Proceedings of the Colorado Scientific Society* 1 (1883–1884): 61–66.

Vickers, William B. *History of the City of Denver, Arapahoe County, and Colorado.* Chicago: O. L. Baskin & Co., 1880.

Vigouroux, George E., ed. *Diary of a Mining Investor.* New York: Quick News Publ. Co., 1910.

Wallihan, Samuel S., and T. O. Bigney. *The Rocky Mountain Directory and Colorado Gazetteer for 1871.* Denver: Samuel S. Wallihan & Co., 1870.

Ward, Roswell. *Henry A. Ward: Museum Builder to America.* Rochester Historical Society Publications, edited by Blake McKelvey, vol. 24. Rochester: Rochester Historical Society, 1948.

Waters, Frank J. *Midas of the Rockies: The Story of Stratton and Cripple Creek.* New York: Covici-Friede, 1937.

Wentworth, Frank L. *Aspen on the Roaring Fork.* Edited by Francis B. Rizzari. Lakewood, Colo.: The Editor, 1950.

Western Colorado Congress. *Western Colorado and Her Resources.* Aspen, Colo.: Aspen Times Print, 1891.

Weston, William. *The Cripple Creek Gold District . . . How It Was Formed. (In Plain English).* Colorado Springs, Colo.: Tilney & Tilney [1896].

Wetherbee, John, Jr. *A Brief Sketch of Colorado Territory and the Gold Mines of that Region.* Boston: Wright & Potter, 1863.

————. *A Letter on Colorado Matters to the Stockholders of Excelsior Co. and others whom it may concern.* Boston: J. E. Farwell & Co., 1867.

————. *A Statement for the Stockholders of the Excelsior, New England, and Invincible Mining Companies, of Colorado.* Boston: Wright & Potter, 1864.

White, Frank M. "The Gold Mines of Colorado." *Harper's Weekly,* June 9, 1894, pp. 538–39.

Whitney, J. Parker. *Colorado, in the United States of America.* London: Cassell, Petter, & Galpin, 1867.

————. *Reminiscences of a Sportsman.* New York: Forest & Stream Publ. Co., 1906.

————. *Silver Mining Regions of Colorado.* New York: D. Van Nostrand, 1865.

Who's Who in New York City and State. Rev. ed. New York: L. R. Hamersly Co., 1905.

Williams, Albert, Jr. "Investments in Mining Properties." *Engineering Magazine* 1 (September, 1891): 731–40.

Wolle, Muriel S. *Stampede to Timberline: Ghost Towns and Mining Camps of Colorado.* Denver: Sage Books, 1949.

Woodard, Bruce A. *Diamonds in the Salt.* Boulder, Colo.: Pruett Press, 1967.

Wray, Henry R., comp. *Some Facts About Cripple Creek, Colorado: America's Greatest Gold Camp.* Colorado Springs, Colo.: Board of Trade and Mining Exchange of Colorado Springs, 1895.

Yeatman, Pope, and Edwin S. Berry. *Mining Securities.* Boston: American Institute of Finance, 1922.

York, C. S., ed. *San Juan's True Fissures.* Vol. 1. Silverton: The Editor, 1880.

Index